Prospects on Cardamom Growers-Issues and Challenges

Dr.S. Thangaraju

Published by

BONFRING®
Intellectual Integrity

Prospects on Cardamom Growers-Issues and Challenges

9 789386 176042 >

ISBN 978-93-86176-04-2

Author

Dr.S. Thangaraju

Bonfring

309, 2nd Floor, 5th Street Extension, Gandhipuram,

Coimbatore-641 012.

Tamilnadu, India.

E-mail: info@bonfring.org | Website: www.bonfring.org

Phone: 0422-3928700

Preface

Cardamom Cultivation and marketing is unique feature among the Plantation crops in India. This book describes about the current scenario on Cardamom Marketing, cultivation and Small Growers role and paradigm shift towards cropping pattern in their Cardamom Estate. India is one among the largest producers of plantation crops in the world. About, three fourth of export earning is derived from agricultural sector and one fourth of earnings derived from oil exports, it clearly therefore indicates that the plantation sector is highly crucial in Indian economy. Despite such features and economic value, cardamom cultivation is less concentration among the Growers especially southern part of India. Obviously we can see the nature and supply and Demand, Cardamom is predominantly taken place the leading in Export. Then, why a number of Growers who unable to maintained their Estates due to chemical inputs. What are the suitable alternate source for which rejuvenate their plants with minimum cost; i.e., organic Farming. Therefore, this book reflect the current situation and remedial measures on Cardamom cultivation and marketing aspects.

Preface

[illegible]

Chapter	Contents	Page No

CHAPTER-I

Introduction

Cardamom, also known as "Queen of Spices" is one of the most exotic and highly priced spices across the globe after Vanilla and Saffron. The most traded varieties of small cardamom are the Malabar and the Guatemalan. Earlier, India was the single largest producer and supplier of Cardamom to the world but, from 2002 onwards, Guatemala replaced India as the premier producer of the spice. Even though India was pushed to second place, it still plays vital role in international trade because of the aroma and flavor of Indian cardamom. Cardamom is also grown in Sri Lanka and Tanzania.

Importance and Area of Plantation Crops

The term Plantation crops refers to those crops which are cultivated on an extensive scale in a large contiguous area, owned and managed by an Individual or a company. The crops include tea, coffee, rubber, cocoa, coconut, arecanut, oil palm, palmyrah, cashew, cinchona etc. These plantation crops are high value commercial crops of greater economic importance and play a vital role in our Indian economy. The main draw back with this sector: of crops in India is that major portion of the area is of small holdings (except Tea) which hinders the adoption of intensive cultivation. In the case of coffee 97.13 per cent of the growers have holdings below ten hectares and in Rubber, 82 per cent of the total area is of small holdings having an average size of 0.5 ha.

The Economic Importance of these Crops are

1. They contribute to national economy by way of export earnings. These crops occupy less than 2 per cent of the total cultivated area (i.e. 3.82 per cent of total crop land) but they generate an income of around Rs. 16,000 million or about 12.72 per cent of the total export earnings of ail commodities or 75 per cent of total earnings from the export of agricultural produces.
2. India is the leading country in the total production of certain plantation crops in the world. For instance, our production meets the share of 47 per cent in tea and 66 per cent in each of cashew and arecanut,
3. Plantation industry provides direct as well as indirect employment lo many millions of people. For instance, tea industry offers direct employment to 10 lakhs and indirect employment to 10 lakh people, while-cashew processing factories alone provide

employment to 3 lakhs people besides 2 lakhs farmers are employed in cashew cultivation.

4. Plantation industry supports many by-product industries and also many rural industries. For example, coconut husk is used to produce coir fiber annually to a tune of 2,19,600 tones in India.
5. These crops help to conserve the soil and ecosystem. Tea planted in hill slopes and cashew in barrel and waste lands protect the land from soil erosion during the rainy season or due to heavy winds.

All India Area, Production and Yield of Cardamom and Pepper

Year	Area in '000 Hectares	Production in Tonnes	Yield in Kg/Hectare
(1)	(2)	(3)	(4)
Cardamom			
1997-98	94.30	11.70	123
1998-99	85.00	9.00	106
1999-2000	84.80	10.70	126
2000-01	92.40	14.50	157
2001-02	91.60	15.20	166
2002-03	88.50	15.40	174
2003-04	95.60	16.70	175
2004-05	95.20	16.90	178
2005-06	95.80	17.80	186
2006-07	98.20	15.70	160
2007-08	81.80	13.40	164
2008-09	91.99	15.45	168
2009-10	90.20	15.72	174
2010-11	87.00	16.00	184
2011-12	89.00	15.80	178
2012-13	92.40	18.40	199
2013-14 (3rd Adv. Est.)	99.20	20.70	209
Pepper			
1997-98	181.50	57.30	316
1998-99	239.80	75.70	316
1999-2000	195.60	59.00	302
2000-01	213.90	63.70	298
2001-02	219.38	62.44	285
2002-03	224.40	71.70	320
2003-04	233.40	73.20	314
2004-05	228.30	73.02	320
2005-06	260.23	92.90	357
2006-07	245.96	69.01	281
2007-08	197.00	47.06	239
2008-09	238.71	47.40	199
2009-10	195.92	51.02	260
2010-11	184.00	52.00	183
2011-12	200.30	40.60	203
2012-13	124.60	52.60	422
2013-14 (3rd Adv. Est.)	121.70	48.70	400

Source: Directorate of Areca nut and Spices Development (DASD)

Estimated Budgetary Requirement for Horticulture Development Programmes of Horticulture Division of DAC and Commodity Boards of the Ministry of Commerce

Estimated Budgetary Requirement for 12th Plan

S. No.	Restructured Programmes	Rs. in Crores
1	Area Expansion Programmes in Integrated Project Mode including protected cultivation	10000
2	For Productivity Enhancement of Existing Orchards / Production Clusters in Project Mode including pollination support	10000
3	Additional Provision for NE region, Hilly areas and scheduled areas and small & marginal Farmers etc for above two schemes	3000
4	Infrastructure development for Input production	250
5	Supply Chain Management, Post Harvest Management	7300
6	Marketing Infrastructure, HRD & Market information	2000
7	Enhancing Export Competitiveness	250
8	Human Resource Development	250
9	Horticulture Database	175
10	Crop Insurance and Setting up Weather Stations	250
11	Schemes of Coconut Development Board	1000
12	Schemes Components of Bamboo Development	500
13	Horticulture Promotion Services- NCCD	25
14	Schemes of Tea Board	1600
15	Schemes of Coffee Board	1200
16	Schemes of Rubber Board	1180
17	Schemes of Spice Board	900
	Total	**39880 crores**

Horticulture & Plantation Sector has received focused attention in our planning process from 7th five year plan period onwards; as a result, there has not only been sustained increase in production of horticulture &plantation crops but hi-tech horticulture has also been recognised as a commercial proposition. It is a fact that horticulture &plantation sector has provided opportunity of crop diversification, resulting into increased income from the land and also the nutrition security.

Plantation and Aromatic-Growth Trends

Background: Plantation crops of tea, coffee, rubber, cardamom, oil palm and medicinal & aromatic plants have been analysed in this chapter. Out of the fore-mentioned crops, tea, coffee, rubber and cardamom are dealt by respective Commodity Boards under Ministry of Commerce; oil palm is dealt by Crops Division of Department of Agriculture& Cooperation, Ministry of Agriculture and medicinal plants by the Department of Ayurveda, Yoga Naturopathy, Unani and Homeopathy, Unani, Siddha and Homeopathy (AYUSH). In view of this fact, growth trend with factors affecting the same have been analysed for making policy recommendations for XIIth plan period.

Tea: Status of the Crop: Currently Tea is grown in about 36 countries spread over 3.3 million hectares. The major tea producing countries are China, India, Sri Lanka, Kenya and Indonesia. These five countries alone account for 78% of world tea production and 75% of exports.

India is the largest producer of black tea as also the largest consumer of tea. It occupies the second position in the world with regard to the area under tea cultivation and is the fourth largest tea exporter in the world. Indian tea contributes to about 26% of the total world tea production.

Tea Plantations in India are mainly located in the hills of North eastern and Southern States. Major tea growing areas of the country are concentrated in Assam, West Bengal, Tamil Nadu and Kerala. The other areas where tea is grown to a small extent are Karnataka, Tripura, Himachal Pradesh, Uttarakhand, Arunachal Pradesh, Manipur, Sikkim, Nagaland, Meghalaya, Mizoram and Bihar. There are about 1,59,190 tea holdings with an area of 5.79 lakh hectares of which 1,57,504 are small holdings having an area less than 10.12 ha. and 1686 big gardens. About 30 lakh workers are employed in the tea sector directly or indirectly.

Small and Large Cardamom

The Department of Commerce through the Spices Board is responsible for the overall development, marketing and export promotion of both small and large cardamom. The details regarding area, production, productivity, exports, imports and consumption of cardamom(small and large) for the last eleven years are as under:

Small Cardamom

			Small Cardamom			
Year	Area (Hects.)	Production (MT)	Productivity (Kgs/Hect)	Exports MT	Import MT	Domestic consumption MT
2000-01	72,320	10,480	195	1,545	88	8,500
2001-02	72,663	11,365	210	1,031	321	10,100
2002-03	73,125	11,920	218	682	323	10,900
2003-04	73,237	11,580	210	757	60	10,300
2004-05	73,725	11,415	206	642	352	10,600
2005-06	73,795	12,540	227	863	437	11,500
2006-07	73,228	11,235	208	650	623	10,600
2007-08	69,300	9,450	191	500	875	9,300
2008-09	71,170	11,000	219	750	180	9,900
2009-10	71,110	10,075	201	1,975	95	7,800
2010-11	71,012	10,380	206	1,175	75	8,800
CAGR	-0.18	-0.1	0.54	-2.7	-1.59	0.35

Large Cardamom

Large Cardamom

Year	Area 000 Ha	Production MT	Productivity Kg/Ha	Export MT	Import MT	Domestic Consumption MT
2000-01	26,358	5,200	232	1,506	4,816	8,510
2001-02	30,008	5,850	239	1,577	4,038	8,311
2002-03	30,008	5,300	209	1,450	4,319	8,169
2003-04	30,039	6,154	246	924	4,141	9,371
2004-05	30,039	5,773	230	954	4,100	8,919
2005-06	30,039	5,185	207	1,046	4,935	9,074
2006-07	30,039	4,480	200	1,500	6,297	9,277
2007-08	30,039	4,920	225	1,325	5,850	9,445
2008-09	27,034	4,300	214	1,875	5,950	8,375
2009-10	27,034	4,180	208	1,000	6,000	9,180
2010-11	26,984	3,918	208	775	4,050	7,193
CAGR	0.23%	-2.79%	-1.09%	-6.43%	-1.72%	-1.67%

Oil Palm- Oil palm is the highest yielder of vegetable oil which was initially introduced in the country in Kerala and Andaman&Nicobar Islands. Subsequently, under the aegies of TMOP&M. Cultivation of oil palm was extended through the centrally sponsored Oil Palm Development Programme (OPDP) to other potential areas in the states of Andhra Pradesh, Karnataka, Tamilnadu, Orissa, Gujarat and Goa since 1992 onwards.

Though the programme was also sanctioned for Assam and Tripura but no tangible progress was made in these two states upto Nineth Plan. Under OPDP, an area of 59229 ha. was planted during the period from 1992 to 2003. However, uprooting of the palms took place in some areas due to sharp decline in domestic edible oil prices due to huge imports of edible oils at cheaper prices during 1999 to 2002 and lack of processing back up in some of the states. To support the oil palm development programmes, following infrastructure and other facilities have already been created:

a. Seed gardens with a production capacity of producing 1.5 million seed sprouts along with nurseries set up by the entrepreneurs.
b. Processing capacity to the extent of 110.5 tonnes FFB per hour (i.e..5.525 lakh tones FFB per annum @ 20 hr. X250 days running).
c. Leaf analysis laboratory.
d. National Research Centre for Oil Palm to provide requisite research support to the crop development programmes.

Medicinal and Aromatic Plants: India has 15 Agro-climatic zones and 17000-18000 species of flowering plants of which 6000-7000 are estimated to have medicinal usage in folk and documented systems of medicine, like Ayurveda, Siddha, Unani and Homoeopathy. About 960 species of medicinal plants are estimated to be in trade of which 178 species have annual consumption levels in excess of 100 metric tones. Medicinal plants are not only a major resource base for the traditional medicine & herbal industry but also provide livelihood and health security to a large segment of Indian population. The domestic trade of the AYUSH industry is of the order of Rs. 80 to 90 billion (1US$ = Rs.50). The Indian medicinal plants and their products also account of exports in the range of Rs. 10 billion. There is global resurgence in traditional and alternative health care systems resulting in world herbal trade which stands at US$ 120 billion and is expected to reach US$ 7 trillion by 2050. Indian share in the world trade, at present, however, is quite low. Not much of data is available about area, production and productivity of medicinal and aromatic plants in the Country.

The National Medicinal Plants Board(NMPB)set-up in November2000 by the Government of India has the primary mandate of coordinating all matters relating to medicinal plants and support policies and programmes for growth of trade, export, conservation and cultivation.

The Board is located in the Department of Ayurveda, Yoga&Naturopathy, Unani, Siddha &Homeopathy (AYUSH) of the Ministry of Health & Family Welfare.

In addition, Ministry of Commerce through Council for Scientific & Industrial Research (CSIR) has set up Central Institute Medicinal and Aromatic Plants at Lucknow with research centres at Bangaluru, Hyderabad, Pantnagar and Purara(Uttarakhand) for varietal development, setting certification standards for seeds, processing, quality control etc.

Targeted Annual Growth Rate for XIIth Five Year Plan Period

a. Targeted Growth Rate for horticulture sector has been worked out with main considerations of nutritional security and our capacity and preparedness to achieve the same keeping in view the SWOT Analysis.
b. Population projections made by Registrar General of Census, Government of India have been taken into account for the purpose of calculating per capita gross availability of fruits&vegetables. However, for the purpose of calculation of per capita net availability of fruits &vegetables, PHM losses to tune of 25% and deduction of 5% for export & processing purpose is applied.
c. With above presumptions in place, it is targeted to achieve recommended level of per capita net availability of fruits and vegetables i.e. 120 gm and 280 gm per capita per

day as recommended by Indian Institute of Nutrition during 12th plan period. For this purpose, growth rate of 6.5 % per annum is targeted for 12th plan period in view of the fact that the Country must have enough surpluses to develop its exports market on sustainable basis.

d. It is noteworthy that our country has recorded a growth rate of 4.4% % in respect of fruits production and 5 % per annum in vegetable production during the year 2010-11 and the same has been taken into account for calculation of base production levels at the beginning of 12th Plan period. However, a higher growth rate of 6% for fruits and 7.5% for vegetables has been targeted during 12th Plan period. If the targeted growth rate is achieved, the country may attain the desired availability level of 120 gms/capita/day of fruits by the end of year 2015-16 and 280 gms/capita/day of vegetables during the year 2016-17

Recommended Course of Action for Supply Chain Management during XIIth Plan Period

Continuation of ongoing schemes of PHM: It is proposed to continue with ongoing schemes relating to Post Harvest Management of horticulture produce. As the cost norms and subsidy norms have been revised recently, same level of assistance may be continued. However, ceiling of 5000 MT put on cold storage capacity need to be relaxed in case of infrastructure under National Green Grid and Terminal and Wholesale Markets.

Development & Transfer of Technology/Technology Solutions: In order to make recommendations of post harvest management, packaging and storages relevant and acceptable to stake-holders it is necessary to develop appropriate technology and validate available technology in our conditions in following respect. Ongoing scheme of Development & Transfer of Technology/Technology under NHB may be continued to address to following needs:

a. Introduction of Production Technology for increasing percentage of quality fruits in a production lot.
b. Development of appropriate design of packaging for horticulture produce, for relevant conditions in which transportation is to be made.
c. Validation of critical storage conditions for Indian cultivars and in our agro-climatic conditions.
d. Validation of PHM protocols for Indian cultivars and in our agro-climatic conditions.

e. Development and Introduction of economically viable Long Distance Bulk Transport Solutions for fresh horticulture produce.

Strengthening of NCCD: National Centre for Cold Chain Development which had been promoted by NHB during XIth Plan Period under its plan scheme need to be strengthened and its ambit may be expanded so as to develop protocols for above mentioned aspects of Development & Transfer of Technology/Technology Solutions.

Introduction of Scheme for Multi-Modal Transport System as Long Distance Bulk Transport Solutions for fresh horticulture produce-a scheme component of promotion of multi-modal transport system for long distance transportation of fresh fruits &vegetables, linking production centres with major markets need to be introduced on assistance pattern of ongoing scheme of Horticulture Park/Common Facility Centre.

Introduction of Scheme for Common Facility of perishable cargo handling at Inland Container Deports, Railway Yards, Dry Ports, Sea Ports, Air Cargo centres, Land Custom Stations etc- this new component too may be operated on assistance pattern of Horticulture Park/Common Facility Centre.

Need for Putting Policy-Framework in Place-Following policy framework need to be put in place for effective supply chain management of fresh horticulture produce.

Introduction of Project Mode of Scheme Implementation: It is proposed that during XIIth plan period, financial assistance for PHM component should be given in integration with new are-a expansion project or in integration with existing orchards in project mode. Thus, components for assistance will be decided keeping in view the system of production, aggregation of farm produce, its transportation to market and trading system in vogue. This policy decision may clash with XIth plan trend of proliferation of Mission Mode programmes for individual components like NMMI, Farm Machinery Mission, Seeds & Planting Material Mission and may be Cold Chain Mission etc. Following items may be given due importance while preparing such a project:

a. Plastic crates/bins and handling system among producer farmers and registered 'Adhatiyas / brokers in APMC markets.
b. Fruit ripening Chambers for banana, mangoes, papaya etc.
c. Mechanised sorting, grading and handling systems.
d. Cold storages with appropriate technical standards and storage conditions.
e. Development of Infrastructure suitable for storing &grading and handling fresh horticulture produce at market centres, Inland Container Deports and

Ports including sea ports,air cargo centres, Land Port Stations. Produce across the Country-At present perishables does not get free passage against city-area entry rules, on check posts etc. There is a need to introduce a system of free passage for fresh horticulture produce. This aspect has been dealt with in details while dealing with issue of Marketing of fresh horticulture produce.

Rationalisation of Cold Storage Licensing System: At present, cold storage licensing is handled by State Government officials from horticulture/marketing directorates. The procedure and provisions regarding licensing revolve around administrative powers and penalty for defaulters but do not prescribe technical conditions which are necessary for securing licenses. There is a need to introduce a uniform protocol and procedure in this regard and, possibly, entrust this task to NCCD.

Development of Export Support Infrastructure within Country as Mandate under Scheme of Horticulture Division of DAC: At present, there is no clarity about role of various agencies in development of suitable export promotion infrastructure within country. As such, pack houses, quality control laboratories, cold chain infrastructures, perishable cargo centres at sea ports and air-ports etc are being supported under schemes of horticulture division of DAC as well as by APEDA. Ministry of Home Affairs too has schemes for development of Land Custom Stations at major inland ports. A policy decision in this respect may be taken that inland infrastructure of this nature will be developed under schemes of horticulture Division of DAC as central sector scheme fully funded bŷ central Estimation of Requirement for Infrastructure Development-It is proposed to estimate requirement of pack houses, cold storages and fruit ripening chambers etc in spite of low level of their prevalence so far. It is targeted to provide sorting & grading/conveyor belt/rope-way facility for about 25% of estimated production of fruits and vegetables. Most of the crops require sorting&grading and packing in about a month's period and facility is generally is used for mono-crop. Taking working of pack-house in two shifts of 8 hours each, one may get about 500 hours of operation. Thus an installed capacity of one lakh TPH may be required. Taking cost norm of about Rs. 10000 per TPH the budgetary requirement at assistance level of 40% of capital cost will come out to be Rs. 400 crores.

Similarly, requirement of additional capacity of cold storage for fruits & vegetables not requiring pre-cooling is taken as 10 Million MT@assistance level of Rs. 2400/MT, cold storage for crops requiring pre-cooling as 2 million MT @ assistance level of Rs. 3200/MT and CA storage as 0.5 Million MT @ assistance level of Rs. 28800/MT, onion storage as 5 million MT @ assistance level of Rs. 1500/MT and introduction of 5000 refrigerated vans @ assistance level of Rs. ten lakh/van and 10 rakes/fleets of about 1000MT capacity per cycle for long distance transport solutions is taken. Lump-sum provision of Rs. 100 crore each is proposed for primary processing and packaging solutions including plastic crates/bins. Similarly, for creating scientific ripening capacity of 5 million MT for 4 day ripening cycle in 250 working days, ripening chambers of total ripening capacity of 80000 MT is required to be created. Taking cost norm of Rs. one lakh/MT the estimated budgetary requirement comes out to be Rs. 800 crores. A budget provision of Rs.25 Crores is proposed for NCCD for carrying out its function of certification of PHM and cold chain infrastructures, HRD and applied R & D function.

Vegetable Processing

Background

Major part of fruits grown in India are consumed as table fruits and therefore, table varieties of fruits get preference in horticulture development programmes. Similarly, fresh vegetables are generally cooked in Indian kitchens each time before the meals in preference to pre-cooked vegetables. That is why; a very small quantity of fresh horticulture produce goes for processing for consumption in the form of processed food. However, processing of fruits and vegetables is assuming significance in planning process for horticulture development as due to change in family profile and increasing urbanisation, more and more household have started preferring ready-to eat, easy to cook type fruits&vegetables. As a result of value-addition the producer farmers get price premium for their horticulture produce. In addition, processing may provide viable solution to the problem of glut in the market due to over-production and also to the problem of transport of fresh horticulture produce from remote production clusters to markets. For example-a number of production clusters of fresh fruits and vegetables in NE and Hilly reasons of the Country are situated in geographically remote areas from where the transport of these perishables to consumption centres results in to huge post harvest losses and their processing has been seen to be a viable solution and therefore, Mission Mode programme for horticulture development in such areas namely, HM-NEHA has a special component Mini-Mission-IV and the scheme sub-component of primary processing figures into NHM and NHB Schemes of horticulture development. Apart from the table varieties of fruits which may be available for processing in above mentioned contingencies, there are a number of other fruits and fruit-varieties which are produced for processing such as pineapples, passion fruits, processing varieties of mangoes, walnut, almond, cashew nut, coconut cocoa, arecanut, 'Makhana (Fox Nut)', white onion. Though, reliable statistics regarding F & V processing units set up in the Country with and without assistance under Government schemes is not available ,it is generally observed that following F & V are commonly processed:

- Green peas
- Tomato for puree, ketch-up, sauces, snacks-dressing etc
- Selected varieties of potatoes for flakes, french fries etc
- White onion for dehydration
- A range of vegetables for cut vegetables, frozen vegetables and pickling
- Certain mango varieties for slices, pulp and pickle making
- Pineapples for slices, tit-bits and juice making
- Grapes varieties for raisin making
- Certain Banana and Jack- fruit varieties for flakes making
- Citrus, strawberry, pineapple, apples, plum, peaches and mixed fruits for jam, jelly and orange marmalade
- Guava and Custard apple for ice-cream
- Apple, Nectarine, Guava, Passion fruits nectarine etc for RTS drinks
- Apricot as dry fruit, peach for peach-halves in sugar syrup; walnut, almond, cashew as dry fruits and value added products
- Spices like Chilly, Ginger, Turmeric, Pepper, Coriander, Cummins, Isabgol, Saffron, Vanilla
- Aromatic and Herbal plants

Issues Relating to Supply Chain/Availability of F & V as Raw Material

APMC Act: Some of the key areas of concern regarding the APMC act are; Market Sector Reforms, Abolition of APMC Act, Permitting Contract Farming and transactions outside the regulated mandis, Private Sector Investment in Marketing Infrastructures, Setting Up of Modern Markets like Safal's Wholesale Market in Bangaluru, Safal Exchange for F&V, Modern Flower Auction Houses in Metros etc have not been able to provide a viable solution to sourcing of quality raw materials for processing, especially for fresh F & V.

Other Operational Issues Affecting Processing of F &V

Contract farming for processing varieties and volatility of market of processed item-issues of aggregation and traceability of a number of processing varieties of F&V can be resolved by adoption of contract farming. But,it is also noted that the processors too are not always keen to enter into contract production due to high volatility of market of processed product. For example-case of white onion.

Working capital for storage of raw material: In a number of cases, the raw material supply is seasonal and is required to be procured in a short time frame. Working capital requirement in such cases is quite high, which is generally not available at a reasonable cost of capital.

Credit Market and Sales Return-Market of process food items is, by and large, a credit market on sales return basis. Even the institutional buyers like five star hotels, air lines etc do purchase these items on credit. Due to demand constraints the manufacturers do not get dealers for sole dealership and one dealer stocks products of more than one manufacturer which ultimately results into sales return for a number of manufacturers. This puts additional financial burden of working capital on manufacturers of processed food items.

Recommendations for 12th Plan Period

Suggested Interventions for Fruits and Vegetables and Consumer Industries

There is a need to prepare a multi-pronged strategy to address all the concerns and challenges identified in the previous section and the following interventions are suggested:

Cataloguing of process able F&V and validated processing technologies: A number of F & V processing projects are not viable ab initio but shown to be viable based on unrealistic presumptions. For example- a project based on processing of such F&V as raw materials which may not be available for processing in estimated quantity due to price premium for table purpose useexcept during periods of over-production or; a project based on such F&V which maynot be process worthy or for processing of which appropriate technology is not within the reach of the entrepreneur. Therefore, rather than promoting investment on projects relating to processing of entire range of F&V produced in a particular area, focus fruits and vegetables with potential for processing, marketable surplus available, and consumer demand in the domestic as well as export markets should be identified. Thus, there is a need that catalogues of F&V which are process-worthy and which is generally available at economically viable price range for processing. Such list of F&V may be termed as "Focus Crops for Processing during 12th Plan Period. Similarly, cataloguing of validated processing technologies may prove to be handy for the entrepreneurs and the same may be named as "Validated Technologies".

Recommendations of the Working Group for 12th Plan

Putting in Place a Unified Technology Support Group:It is proposed that central schemes of horticulture development; namely, schemes of NHB, NHM, HMNEH, NMMI and NBM may be integrated. For effective transfer of technology, there is a need to have a Unified Technology Support Group for the Integrated National Horticulture Programme which will be responsible for ensuring proper identification and transfer of technology solutions to field level issues relating to horticulture projects. This Group will work in close coordination with ICAR Institutions and SAUs on one side and private sector R&D and International Technical Collaboration Programmes on the other and ensure consolidation of appropriate extension messages and preparation of technology intervention plans.

Similarly, Director(Agriculture/Horticulture) should lead Technology Support Group at level of State Mission Directorates and ensure consolidation of appropriate extension messages and preparation of technology intervention plans and transfer of the same to extension workers in selected system of extension for finally passing on the same to the producer farmers.

De-fragmentation of Agriculture Extension System: Vertical bifurcation of agriculture extension system into two-agriculture and horticulture has been affected in most of the States with stated objectives of developing clear-cut line of command under two separate directorates of agriculture and horticulture and promoting development of sectoral expertise. This has not only resulted into thinning of the two streams of extension but also does not actually serve purpose of effective extension and conflicts with the ground reality that a farmer needs integrated advise for all the crops grown by him. It is proposed that States may consider introducing single-window extension system for agriculture, horticulture, soil & water-conservation etc.

Cluster based extension for targeted deliverables: There is also a need to move away from mode of extending component-wise assistance to project-mode of extension and accordingly, extension message for every target cluster may be worked out in details; the same may be transferred to the producer farmers in the selected cluster. For this purpose, it is also necessary to set out deliverables of technology interventions such as increased productivity, increase in percentage of quality fruits, IPM, organic farming, GAP certification, PHM practices etc and extension messages should be designed accordingly with practical tips. Beneficiaries for new projects should be imparted training during project implementation.

The entire exercise may be carried out with the help of organizations selected carefully on the basis of their capacity and interest.

Area of Plantation Crops in India

The area and production is more in Kerala (1073000.7 ha and 4359000.9 MT) followed by Karnataka (679000.2 ha and 1401000.0 MT), Tamil Nadu (503000.0 ha and 3810.6 MT), Andhra Pradesh (289000.0 ha/and 1021000.2 MT).

CHAPTER-II

Agriculture Policy: Vision 2020

India has made impressive strides on the agricultural front during the last three decades. Much of the credit for this success should go to the several million small farming families that form the back bone of Indian agriculture and economy. Policy support, production strategies, public investment in infrastructure, research and extension for crop, livestock and fisheries have significantly helped to increase food production and its availability. During the last 30 years, India's food grain production nearly doubled from 102 million tons in the triennium ending 1973 to nearly 200 million tons (mt) in the triennium ending (TE) 1999. Virtually all of the increase in the production resulted from yield gains rather than expansion of cultivated area. Availability of food grains per person increased from 452 gm/capita/day to over 476 gm/capita/day, even as the country's population almost doubled, swelling from 548 million to nearly 1000 million.

Increased agricultural productivity and rapid industrial growth in the recent years have contributed to a significant reduction in poverty level, from 55 percent in 1973 to 26 percent 1998. Despite the impressive growth and development, India is still home to the largest number of poor people of the world. With about 250 million below the poverty line, India accounts for about one-fifth of the world's poor. Child malnutrition extracts its highest toll in this country. About 25% children suffer from serious malnutrition. More than 50 percent of the pre-school children and pregnant women are anemic. The depth of hunger among the undernourished is also high. India has high population pressure on land and other resources to meet its food and development needs. The natural resource base of land, water and bio-diversity is under severe pressure. The massive increase in population (despite the slowing down of the rate of growth) and substantial income growth, demand an extra about 2.5 mt of food grains annually, besides significant increases needed in the supply of livestock, fish and horticultural products. Under the assumption of 3.5% growth in per capita GDP (low income growth scenario), demand for food grains (including feed, seed, wastage and export) is projected in the year 2020 at the level of 256 mt comprising 112mt of rice, 82mt of wheat, 39mt of coarse grains and 22mt of pulses. The demand for sugar, fruits, vegetables, and milk is estimated to grow to a level 33mt, 77mt, 136mt and 116mt respectively. The demand for meat is projected at 9mt, fish 11mt and eggs 77.5 billion Future increases in the production of cereals and non cereal agricultural commodities will have to be essentially achieved through increases in productivity, as the possibilities of expansion of area and livestock population are minimal. To meet the projected demand in the year 2020, country must attain a per hectare yield of 2.7 tons for rice, 3.1 tons for wheat, 2.1 tons for maize, 1.3 tons for coarse cereals, 2.4 tons for cereal, 1.3 tons for pulses, 22.3 tons for potato, 25.7 for vegetables, and 24.1 tons for fruits. The production of livestock and poultry products must be improved 61% for milk, 76% for meat, 91% for fish, and 169% for eggs by the year 2020 over the base year TE 1999. Average yields of most crops in India are still rather low.

Emerging Trends The agriculture sector recorded satisfactory growth due to improved technology, irrigation, inputs and pricing policies. Livestock, poultry, fisheries and horticulture are surging ahead in production growth in recent years and will have greater demand in the future. Industrial and service sectors have expanded faster than agriculture sector resulting in declining share of agriculture in national accounts. Despite the structural change, agriculture still remains a key sector, providing both employment and livelihood opportunities to more than 70 percent of the country's population who live in rural areas. The contribution of small farmers to the national and household food security has been steadily increasing. The water availability for agricultural uses has reached a critical level and deserves urgent attention of all concerned.

India has high population pressure on land and other resources to meet its food and development needs. The natural resource base of land, water and bio-diversity is under severe pressure. Food demand challenges ahead are formidable considering the non-availability of favourable factors of past growth, fast declining factor productivity in major cropping systems and rapidly shrinking resource base. Vast uncommon opportunities to harness agricultural potential still remain, which can be tapped to achieve future targets. There are serious gaps both in yield potential and technology transfer as the national average yields of most of the commodities are low, which if addressed properly could be harnessed. Concentration was on enhanced production of a few commodities like rice and wheat, which could quickly contribute to increased total food and agricultural production. This resulted in considerable depletion of natural resources and the rained dry areas having maximum concentration of resource poor farmers remained ignored, aggravating problems of inequity and regional imbalances. This also led to a high concentration of malnourished people in these rainfed, low productive areas. This era also witnessed rapid loss of soil nutrients, agro-biodiversity including indigenous land races and breeds. The agriculture policy must accelerate all-round development and economic viability of agriculture in comprehensive terms. Farmers must be provided the necessary support, encouragement and incentives. It must focus both on income and greater on-farm and off-farm job and livelihood opportunities.

Main Issues

In national priority setting, the following recurring and emerging issues for sustainable agricultural development and poverty alleviation must be considered:

- Population pressure and demographic transition;
- Resource base degradation and water scarcity;
- Investment in agriculture, structural adjustment and impact on the poor;
- Globalization and implication on the poor;
- Modern science and technology and support to research and technology development; and
- Rapid urbanization and urbanization of poverty, and deceleration in rural poverty reduction.

In addressing the above issues, a policy statement on agriculture must take note of the following uncommon opportunities:

- Conservation of natural resources and protection of environment.
- Vast untapped potential of our soil and water resources, and farming systems
- Technology revolution especially in the areas of molecular biology, biotechnology, space technology, ecology and management.
- Revolution in informatics and communication and the opportunity of linking farmers, extension workers and scientists with the national and international databases Vision. The Agriculture Policy document must articulate a clear vision on following few basic parameters of the agricultural sector around which a policy framework must be developed.
- Organization of agriculture: A clear long-term vision where inter-sectoral linkages are explicit.
- Sustainability and natural resource management: Prescription must lie in the domain of political economy.Otherwise, allocating funds for watershed development, agro forestry, soil conservation, and so on will not produce desired results.
- Institutional change: Policy document must spell out new approaches and new institutions free from the shackles of bureaucratic and self-help framework.
- Investment priorities: There is a need to develop a consensus on investment themes, priorities and policies. Policy document must lend strength to the claim for greater investment in rural areas, and also re-examine its programmes in the light of complementarities.

- Incentives: Document must articulate a clear vision on the incentive framework.

Risk management Challenges,Policies and Strategies Enhancing Yield of Major Commodities Yield of major crops and livestock in the region is much lower than that in the rest of the world. Considering that the frontiers of expansion of cultivated area are almost closed in the region, the future increase in food production to meet the continuing high demand must come from increase in yield. There is a need to strengthen adaptive research and technology assessment, refinement and transfer capabilities of the country so that the existing wide technology transfer gaps are bridged. For this, an appropriate network of extension service needs to be created to stimulate and encourage both top-down and bottom-up flows of information between farmers, extension workers, and research scientists to promote the generation, adoption, and evaluation of location specific farm technologies. Ample scope exists for increasing genetic yield potential of a large number of vegetables, fruits as well as other food crops and livestock and fisheries products. Besides maintenance breeding, greater effort should be made towards developing hybrid varieties as well as varieties suitable for export purposes. Agronomic and soil researches in the region need to be intensified to address location specific problems as factor productivity growth is decelerating in major production regimes. Research on coarse grains, pulses and oilseeds must achieve a production breakthrough. Hybrid rice, single cross hybrids of maize and pigeonpea hybrids offer new opportunities. Soybean, sunflower and oil palm will help in meeting future oil demands successfully. Forest cover must be preserved to keep off climatic disturbances and to provide enough of fuel and fodder. Milk, meat and draught capacity of our animals needs to be improved quickly through better management practices. Integrated nutrient management: Attention should be given to balanced use of nutrients. Phosphorus deficiency is now the most widespread soil fertility problem in both irrigated and unirrigated areas. Correcting the distortion in relative prices of primary fertilizers could help correct the imbalances in the use of primary plant nutrients -nitrogen, phosphorus, and potash and use of bio-fertilizers. To improve efficiency of fertilizer use, what is really needed is enhanced location-specific research on efficient fertilizer practices (such as balanced use of nutrients, correct timing and placement of fertilizers, and, wherever necessary, use of micronutrient and soil amendments), improvement in soil testing services, development of improved fertilizer supply and distribution systems, and development of physical and institutional infrastructure. Arresting deceleration in total factor productivity:

Public investment in irrigation, infrastructure development (road, electricity), research and extension and efficient use of water and plant nutrients are the dominant sources of TFP growth. The sharp deceleration in total investment and more so in public sector investment in agriculture is the main cause for the deceleration. This has resulted in the slow-down in the growth of irrigated area and a sharp deceleration in the rate of growth of fertilizer consumption.

The most serious effect of deceleration in total investment has been on agricultural research and extension. This trend must be reversed as the projected increase in food and non-food production must accrue essentially through increasing yield per hectare. Recognizing that there are serious yield gaps and there are already proven paths for increasing productivity, it is very important for India to maintain a steady growth rate in total factor productivity. As the TFP increases, the cost of production decreases and the prices also decrease and stabilise. Both producer and consumer share the benefits.

The fall in food prices will benefit the urban and rural poor more than the upper income groups, because the former spend a much larger proportion of their income on cereals than the latter. All the efforts need to be concentrated on accelerating growth in TFP, whilst conserving natural resources and promoting ecological integrity of agricultural system. More than half of the required growth in yield to meet the target of demand must be met from research efforts by developing location specific and low input use technologies with the emphasis on the regions where the current yields are below the required national average yield. Literacy had a positive and significant relation with crop productivity and a strong link exists between literacy and farm modernisation. A recent study (Kumar and Mittal, 2000), has shown that literacy emerged as an important source of growth in adoption of technology, use of modern inputs like machines, fertilisers, and yield. Recognising that in the liberalised economic environment, efficiency and growth orientation will attract maximum attention. Literacy will play a far more important role in the globalised world than it did in the past. Contribution of literacy, through TFP, will be substantial on yield growth and domestic supply. As future agriculture will increasingly be science-led and will require modern economic management, high return to investment on education is expected. The investments that are good for agricultural growth-technology and its dissemination, rural infrastructure (roads), education and irrigation - amount to a 'win-win' strategy for reducing rural poverty by also increasing the non-farm economy and raising rural wages. Creating infrastructure in less developed areas, better management of infrastructure and introduction of new technologies can further enhance resource productivity and TFP. Generation and effective assessment and diffusion of packages of appropriate technologies involving system and programme based approach, participatory mechanisms, greater congruency between productivity and sustainability through integrated pest management and integrated soil-water-irrigation-nutrient management, should be aggressively promoted to bridge the yield gaps in most field crops. Besides this, efforts must be in place to defend the gains and to make new gains particularly through the congruence of gene revolution, informatics revolution, management revolution and eco-technology.

Many observers have expressed concern that technological gains have not occurred in a number of crops, notably coarse cereals, pulses and in rainfed areas. Recent analysis on TFP growth based on cost of cultivation data does not prove this perception. In all the 18 major crops considered in the analysis, several states have recorded positive TFP growth. This is spread over major cereals, coarse grains, pulses, oilseeds, fibres, vegetables, etc. In most cases, in the major producing states, rainfed crops also, showed productivity gains. There is thus strong evidence that technological change has generally pervaded the entire crop sector. There are, of course, crops and states where technological stagnation or decline is apparent and these are the priorities for present and future agricultural research. Farming system research to develop location specific technologies and strategy to make grey areas green by adopting three-pronged approach-watershed management, hybrid technology and small farm mechanization will accelerate growth in TFP. It is necessary to enlarge the efforts for promoting available dry land technologies. Promoting efficient fertiliser practices, improving soil-testing services, strengthening distribution channel of critical inputs specially quality seed and development of physical and institutional infrastructure will help resource-poor farmers.

Bridging Yield Gaps

Vast untapped potential in the yield exists for all crops in most of the states accounting for more than three-fourths of crop area. Emphasis must be given to the states in which current yield levels are below the national average yield.

Bihar, Orissa, Assam, West Bengal and Uttar Pradesh are the priority states accounting for 66% of rice area which need emphasis on bridging yield gaps to attain target demand and yield growth. For wheat we must focus mainly on Uttar Pradsh, Madhya Pradesh, Bihar and Rajasthan accounting for 68% of wheat area. For coarse cereals, major emphasis must be given to Rajasthan, Maharashtra, Karnataka, Madhya Pradesh, Andhra Pradesh and Uttar Pradesh. To meet the demand for pulses greater emphasis is needed in almost all the states with particular focus on Madhya Pradesh, Maharashtra, Rajasthan, Gujarat, Andhra Pradesh, Karnataka and Uttar pradesh which have three-fourths of total pulse area. The target growth in pulse yield from these states annually must be 6 per cent; otherwise the nation will experience shortage of pulses for all times to come. The task of attaining self sufficient in pulses production looks difficult without area expansion and irrigation. In cases of oilseeds greater emphasis is needed on Andhra Pradesh, Madhya Pradesh, Rajasthan, Maharashtra, Karnataka, West Bengal and Uttar Pradesh to increase the yield by about 4 per cent. The possibilities of developing processing industry for extracting edible oils from non-oilseeds commodities, like rice bran etc, needing to be explored. The introduction of palm cultivation for oil production may release pressure on traditional oilseeds crops to meet future edible oil demand. In case of sugarcane, research and development efforts are to be strengthened in Uttar Pradesh and Bihar to increase the yields per hectare by about 4% per annum. The demand for sugar can also be met by developing mini sugar mills so that substantial sugarcane production can be diverted from Khandsari to sugar production. This may also help release some sugarcane area to other crops. Cotton crop requires greater yield improvement emphasis on 81 per cent of the cotton area in Maharashtra, Gujarat and Andhra Pradesh.

Water for Sustainable Food Security India will be required to produce more and more from less and less land and water resources. Alarming rates of ground water depletion and serious environmental and social problems of some of the major irrigation projects on one hand, and the multiple benefits of irrigation water in enhancing production and productivity, food security, poverty alleviation, as mentioned earlier, are well known to be further elaborated here: In India, water availability per capita was over 5000 cubic metres (m) per annum in 1950. It now stands at around 2000 m and is projected to decline to 1500 m by 2025. Further, the quality of available water is deteriorating. Also, there are gross inequalities between basins and geographic regions.

Agriculture is the biggest user of water, accounting for about 80 percent of the water withdrawals. There are pressures for diverting water from agriculture to other sectors. A study has warned that re-allocation of water out of agriculture can have a dramatic impact on global food markets. It is projected that availability of water for agricultural use in India may be reduced by 21 percent by 2020, resulting: in drop of yields of irrigated crops, especially rice, thus price rise and withdrawal of food from poor masses. Policy reforms are needed from now to avoid the negative developments in the years to come. These reforms may include the establishment of secure water rights to users, the decentralization and privatization of water. management functions to appropriate levels, pricing reforms, markets in tradable property rights, and the introduction of appropriate water-saving technologies. The needs of other sectors for water cannot be ignored. Therefore it is necessary that an integrated water use policy is formulated and judiciously implemented. Several international initiatives on this aspect have been taken in recent years. India should critically examine these initiatives and develop its country-specific system for judicious and integrated use and management of water. A national institution should be established to assess the various issues, regulatory concerns, water laws and legislations, research and technology development and dissemination, social mobilization and participatory and community involvement, including gender and equity concerns and economic aspects. This institution should function in a trusteeship mode and seen as the flagship of a national system for sustainable water security. Emphasis on Rainfed Ecosystem Resource-poor farmers in the rainfed ecosystems practice less-intensive agriculture, and since their incomes depend on local agriculture, they benefit little from increased food production in irrigated areas. To help them, efforts must be increased to disseminate available dry land technologies and to generate new ones. It will be necessary to enlarge the efforts for promoting available dry land technologies, increasing the stock of this knowledge, and removing pro-irrigation biases in public investment and expenditure, as well as credit flows, for technology-based agricultural growth. Watershed development for raising yields of rainfed crops and widening of seed revolution to cover oilseeds, pulses, fruits and vegetables. Farming system research to develop location specific technologies must be intensified in the rainfed areas. Strategy to make grey areas green will lead to second Green Revolution, which would demand three pronged strategy - watershed management, hybrid technology and small farm mechanisation.

Accent on Diversification of Agriculture and Value Addition In the face of shrinking natural resources and ever increasing demand for larger food and agricultural production arising due to high population and income growths, agricultural intensification is the main course of future growth of agriculture in the region. Research for product diversification should be yet another important area. Besides developing technologies for promoting intensification, the country must give greater attention to the development of technologies that will facilitate agricultural diversification particularly towards intensive production of fruits, vegetables, flowers and other high value crops that are expected to increase income growth and generate effective demand for food. The per capita availability of arable land is quite low and declining over time. Diversification towards these high value and labour intensive commodities can provide adequate income and employment to the farmers dependent on small size of farms. Due importance should be given to quality and nutritional aspects. High attention should be given to develop post-harvest handling and agro-processing and value addition technologies not only to reduce the heavy post-harvest losses and also improve quality through proper storage, packaging, handling and transport. The role of biotechnology in post-harvest management and value addition deserves to be enhanced. Accent on Post-Harvest Management, Value Addition and Cost-Effectiveness Post-harvest losses generally range from 5 to 10 percent for non-perishables and about 30 percent for perishables. This loss could be and must be minimized. Let us remember, a grain saved is a grain produced. Emphasis should therefore be placed to develop post-harvest handling, agro-processing and value-addition technologies not only to prevent the high losses, but also to improve quality through proper storage, packaging, handling and transport. With the thrust on globalization and increasing competitiveness, this approach will improve the agricultural export contribution of India, which is proportionately extremely low. Cost-effectiveness in production and post-harvest handling through the application of latest technologies will be a necessity. The agro-processing facilities should preferably be located close to the points of production in rural areas, which will greatly promote off-farm employment. Such centres of processing and value addition will encourage production by masses against mass production in factories located in urban areas. Agricultural cooperatives and Gram Panchayats must play a leading role in this effort. In doing so, the needs of small farmers should be kept in mind. Increased Investment in Agriculture and Infrastructures. The public investment in agriculture has been declining and is one of the main reasons behind the declining productivity and low capital formation in the agriculture sector. With the burden on productivity-driven growth in the future, this worrisome trend must be reversed. Private investment in agriculture has also been slow and must be stimulated through appropriate policies. Considering that nearly 70 percent of India still lives in villages, agricultural growth will continue to be the engine of broad-based economic growth and development as well as of natural resource conservation, leave alone food security and poverty alleviation.

Accelerated investment are needed to facilitate agricultural and rural development through:

Productivity increasing varieties of crops, breeds of livestock, strains of microbes and efficient packages of technologies, particularly those for land and water management, for obviating biotic, a biotic, socio-economic and environmental constraints;

- Yield increasing and environmentally-friendly production and post-harvest and value-addition technologies;
- Reliable and timely availability of quality inputs at reasonable prices, institutional and credit supports, especially for small and resource-poor farmers, and support to land and water resources development;

- Effective and credible technology, procurement, assessment and transfer and extension system involving appropriate linkages and partnerships; again with an emphasis on reaching the small farmers;
- Improved institutional and credit support and increased rural employment opportunities, including those through creating agriculture-based rural agro-processing and agro-industries, improved rural infrastructures, including access to information, and effective markets, farm to market roads and related infrastructure;
- Particular attention to the needs and participation of women farmers; and
- Primary education,healthcare, clean drinking water, safe sanitation, adequate nutrition, particularly for children (including through mid-day meal at schools) and women. The above investments will need to be supported through appropriate policies that do not discriminate against agriculture and the rural poor. Given the increasing role of small farmers in food security and poverty alleviation, development efforts must be geared to meet the needs and potential of such farmers through their active participation in the growth process. Government should facilitate and support community level action by private voluntary organizations, including farmers groups aimed at improving food security, reducing poverty, and assuring sustainability in the management of natural resources.

In addition, governments should enhance efforts to ensure good nutrition and access to sufficient food for all through primary health care and education for all. Modern biotechnology tools, genetic engineering, as well as conventional breeding methods are all expected to play important roles in the generation of higher yielding, pest and stress resistant varieties of rice, wheat, maize and other cereal crops. The availability of genetic innovations in developing countries will depend on continued high levels of investments in agricultural research, both at the international and the national levels. Free and unhindered access to germplasm to breeders worldwide is absolutely crucial to the rapid dissemination and adoption of improved germplasm. This free movement and the dissemination of modern biotechnology innovations to developing countries are hampered by increased patent protection and private sector investments. There is an urgent need to address this problem of free access to technology in the future. Increased attention will also have to be given to development of sustainable systems that protect the natural resource base. Recent evidence of resource degradation and declining productivity in some intensively cropped areas is of particular concern. Also population driven intensification of agriculture without the use of external inputs, is leading to a serious problem of mining soil fertility Sustaining global food supplies will depend on continued high levels of investments in research and technology development. It is essential that research capacity has to be increased substantially. In addition to investments in research, infrastructure investments, particularly in irrigation, transport and market infrastructure development are equally important for sustaining the productivity and profitability of food crop production. Mobilize the best of science and development efforts (including traditional knowledge and modern scientific approach) through partnerships involving national and international research institutions, NGOs, farmers' organizations and private sector in order to tackle the present and future problems of food security and production. Donors and Government must urgently increase funding for agricultural research targeted at the needs of the rural and urban poor, and every effort must be made to ensure the free flow of information, technology and germplasm so that a proper sustainable agriculture can be achieved.

Fighting Poverty and Hunger Nearly one-fourth of India's population, 251 million out of nearly one billion, is below the poverty line. One hundred seventy millions of the poor, 68 percent, are rural and the remaining 32 percent are urban (Table 4). Number at the national level in rural area has decreased after 1983; the number of poor in the cities has been increasing. This is essentially due to migration of the destitute from villages to cities. There are serious implications of this trend on feeding the cities and food security of urban people, urban poverty and environment. A question may be asked as to whether the rural settings and opportunities could be improved for securing livelihood security and consequently rationalizing the migration to the cities.

An analysis of the incidence of rural poverty and hunger by farm size revealed that more than half of the landless people are poor. Poverty got significantly reduced from 54 percent in the landless group to 38 percent in the sub-marginal group suggesting that even a small piece of land, less than 1/2 hectare, can greatly reduce both poverty and hunger1. The incidence of hunger and poverty gets reduced as one is able to meet even part of his/her dietary energy requirement through growing his/her own food .Studies show that even a small plot of one's own helps women to escape extreme poverty and deprivation. Land is the main asset for livelihood security. Although several factors affect the extent and depth of poverty and hunger, some of them have overwhelming impacts under the Indian setting. These include, irrigation, farming system and literacy. Generally, there is higher concentration of poor, and hungry people in rain fed areas as compared with those in irrigated zones. Even with 20 percent of the irrigation intensity, there is a sharp fall in the proportion of hunger and poverty and it remains there irrespective of further intensification of irrigation.

Evidences suggest that extensive irrigation will prove much more effective than to adding more and more water, and often wasting it along with the associated degradation of the natural resources. Such a policy will not only reduce poverty and hunger, but will also promote equity and environmental protection and natural resource conservation. An effective water policy and institutional support is needed to ensure judicious and equitable allocation, distribution and exploitation of water and water resources.

Livestock sector should also receive high priority with multiple objectives of diversifying agriculture, raising income and meeting the nutritional security of the poor farm households. Literacy has a very high impact on poverty alleviation as well as on hunger reduction. The illiterate people, whether urban or rural, are the most poor and malnourished. In urban areas the impact of literacy on poverty is the highest. Education, even above primary level, is extremely effective in reducing both poverty and hunger. Graduate and technical education is, of course, the most important instrument for reducing both poverty and hunger. But its impact is most visible on poverty reduction. Therefore, the education policy of the country must be geared to remove illiteracy as soon as possible, as 50 percent of our people are still illiterate. Free education up to 8'th standard coupled with mid-day meal in the schools will go a long way in reducing both poverty and hunger and will thus help build a strong India.

Further, this move will greatly reduce the violation of child labour laws and will offset some of the non-tariff restrictions imposed by developed countries on exports from developing countries on the grounds of use of child labour.

Accent on Empowering the Small Farmers Contributions of small holders in securing food for growing population have increased considerably even though they are most insecure and vulnerable group in the society. The off-farm and non-farm employment opportunities can play an important role. Against expectation under the liberalized scenario, the non-agricultural employment in rural areas has not improved. Greater emphasis needs to be placed on non-farm employment and appropriate budgetary allocations and rural credit through banking systems should be in place to promote appropriate rural enterprises. Specific human resource and skill development programmes to train them will make them better decision-makers and highly productive. Human resource development for increasing productivity of these small holders should get high priority. Thus, knowledge and skill development of rural people both in agriculture and non-agriculture sectors is essential for achieving economic and social goals. A careful balance will therefore need to be maintained between the agricultural and non-agricultural employment and farm and non-farm economy, as the two sectors are closely inter-connected.

Raising agricultural productivity requires continuing investments in human resource development, agricultural research and development, improved information and extension, market, roads and related infrastructure development and efficient small-scale, farmer-controlled irrigation technologies, and custom hiring services. Such investments would give small farmers the options and flexibility to adjust and respond to market conditions. For poor farm-households whose major endowment is its labour force, economic growth with equity will give increased entitlement by offering favourable markets for its products and more employment opportunities. Economic growth if not managed suitably, can lead to growing inequalities. Agrarian reforms to alleviate unequal access to land, compounded by unequal access to water, credit, knowledge and markets, have not only rectified income distribution but also resulted in sharp increases in productivity and hence need to be adopted widely. Further, targeted measures that not only address the immediate food and health care requirements of disadvantaged groups, but also provide them with developmental means, like access to inputs, infrastructure, services and most important, education should be taken. Identification of need-based productive programs is very critical, which can be explored through characterisation of production environment. We have to develop demand-driven and location-specific programs to meet the requirements of different regions to meet the nutritional security of most vulnerable population in the rural areas. Improved agricultural technology, irrigation, livestock sector and literacy will be most important instruments for improving the nutritional security of the farm-households. Watershed development and water saving techniques will have far reaching implications in increasing agricultural production and raising calorie intake in the rainfed areas. Livestock sector should receive high priority with multiple objectives of diversifying agriculture, raising income and meeting the nutritional security of the poor farm households. Need based and location-specific community programs, which promise to raise nutritional security, should be identified and effectively implemented. Expansion of micro credit programmes for income-generation activities, innovative approaches to promote family planning and providing primary health services to people and livestock and education should enhance labour productivity and adoption of new technologies. Development of the post-harvest sector, co-operatives, roads, education, and research and development should be an investment priority. A congenial policy environment is needed to enable smaller holders to take the advantage of available techniques of production, which can generate more incomes and employment in villages. For this poor farmer needs the support of necessary services in the form of backward and forward linkages. Small-mechanised tools, which minimise do not reduce employment, but only add value to the working hours are needed to enhance labour productivity. Special safety nets should be designed and implemented for them. Can agricultural co-operatives internalise and galvanize these marginal and excluded people? Off-farm employment provided through co-operatives will go a long way in pulling them out of the state where poverty breeds poverty. Therefore, investment in the empowerment of the small landholders will pay off handsomely. Let us create rural centres of production and processing by massesthrough co-operatives or empowerment of Gram Panchayats to promote co-operatives. This will improve efficiency of input and output marketing and give higher income. There is need to disseminate widely post-harvest handling and agro-processing and value addition technologies not only to reduce the heavy post-harvest losses but also improve quality through proper storage, packaging, handling and transport. Panchayati Raj institutions and co-operatives can play significant role in all these directions. Giving them power over the administration, ascontemplated under the 73rdand 74thAmendment of the Constitution has not been implemented seriously so far in any of the states.

Disaster Management

The frequency and intensity of disasters such as floods, droughts, cyclones and earthquakes have increased in the recent years. The devastating earthquake in Gujarat has brought untold miseries to the whole state and caused a national disaster. Special effort should be made to develop appropriate technologies for increasing preparedness to predict and to manage the disasters.Effective and reliable information and communication systems, contingency planning and national and international mobilization of technologies and resources are a must. Experiences of other countries in prevention and management of the disasters should be shared. Keeping Pace with Globalisation The globalization of agricultural trade will bring to the fore access to markets; new opportunities for employment and income generation; productivity gains and increased flow of investments into sustainable agriculture and rural development. However, this will only come to pass if we are mindful of the interests of billions of small and subsistence oriented farmers, fisher-folk and forest dwellers in the short and medium tern. So far the magic of globalization has not been felt in India. During the past one-decade of liberalization certain trends such as deceleration of the growth rate of agricultural GDP, declaration in yield growth rates, and low non-agricultural employment have emerged against expectations. As we globalize, however, it is imperative that we do not forget social aspirations for a more just, equitable and sustainable way of life. Trade agreements must be accompanied by operationally effective measures to ease the adjustment process for a small farmer in developing countries. Exploiting Cyberspace Information is power and will underpin future progress and prosperity. Efforts must be made to strengthen the informatics in agriculture by developing new databases, linking databases with international databases and adding value to information to facilitate decision making at various levels. Development of production models for various agro-ecological regimes to forecast the, production potential should assume greater importance. Using the remote sensing and GIS technologies, natural and other agricultural resource should be mapped at micro and macro levels and effectively used for land and water use planning as well as agricultural forecasting, market intelligence and e-business, contingency planning and prediction of disease and pest incidences.

Sustainable Practices in Cardamom: An Industrial Perspective

There are two major origins for Cardamom cultivation today: Guatemala in Central America and Southern India (in the order of annual crop quantity). Guatemala, the market leader has a big advantage in export markets because its climate allows for production all the year round, compared to being at the whims and fantasies of the monsoon in India.

India was once the world export leader with an estimated 56 percent of the global market in 1970.

India's share of the world market has declined due to:

1. **High production cost:** Cardamom is a highly labor intensive crop and account for 60-70% of the total production cost.
   The trick is in the picking. Labor shortage is a problem Indian agriculture has been grappling with.
2. **Low yields and decreasing volatile oil (VO) content:** The flavor and aroma profile of Cardamom are characteristic of the volatile oil content in it. The challenge here is to increase total VO content, without increasing the concentration of 1, 8-Cineol - too much of which alters the flavor profile, unfavorably.
3. **Pesticide Residues:** High pesticide use in conventional farming, make the crop unfit for export, because they cannot meet the stringent regulatory norms set by importing countries.
4. **Exportable surplus:** India exports 35% of its production today, compared to around 70% in the 1970s - The percentage exported v/s what is consumed has been going down. Considering the high domestic consumption rate which has been increasing over the years, the creation of exportable surplus in the coming years, would become a challenge.

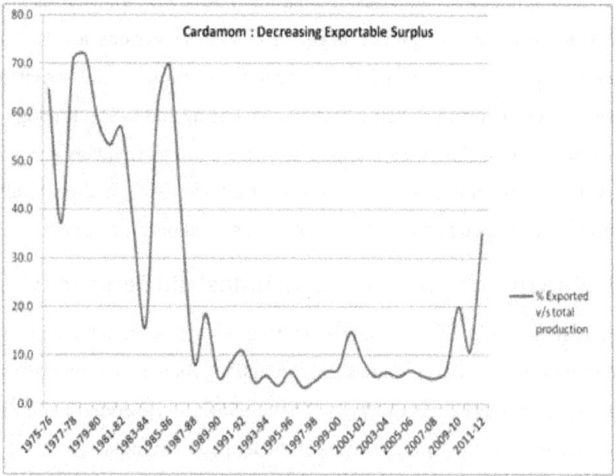

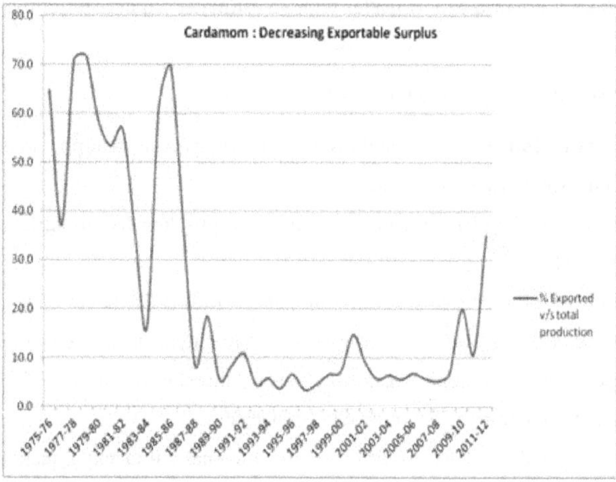

1. *The Increasing Demand for Cardamom*

1. Saudi Arabia is the world's largest importer of Cardamom. A concoction made of Coffee and Cardamom is a popular drink in the Middle East.
2. Cardamom is used in Dutch and Scandinavian cakes and pastries, in mulled wines, and occasionally in Russian liqueurs.
3. Domestic consumption has been steadily on the rise and India continues to be the largest consumer market for Cardamom. It features heavily in curries, pickles, and spice blends (e.g. Garam Masala) in India.

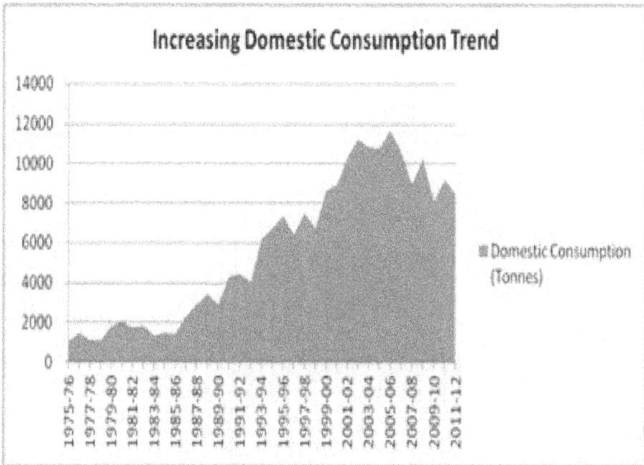

4. **Popular functional health benefits :** Cardamom was used extensively in traditional Chinese and Indian Medicine, for our 3000 years, especially for pulmonary disease, fever, digestive and urinary complications. Hippocrates recommended it for sciatica, coughs, abdominal pains, spasms, nervous disorders, retention of urine and also for bites of venomous creatures.

   Ancient claims are today sanctioned by science and used by Global companies looking for exotic cures to differentiate their products.

2. Moving Up the Value Chain

Cardamom is also used in processed food and cosmetics, as flavors and fragrances, in the form of extracts, oleoresins and essential oils – through steam distillation and solvent extraction of the dried ripe fruit. The principal constituents include Terpinyl acetate and cineol, limonene, sabiene, linalol, linalyl acetate, pinene, zingiberene, among others.

1. **Changing Tastes and Preferences:** The western world due to its geographical proximity to Guatemala has grown accustomed to the flavor profile characteristic of Cardamom from that origin. Crop failures at Guatemala, have had multinational companies like the Wrigley's looking out for alternate origins. They have played a part in shifting global consumer tastes and preference to accept the flavor nuances of the Indian Cardamom.
2. **Higher Productivity:** The Indian area under cardamom cultivation was 75,000 hectares in 1975-76 giving a total production of 3,000 tones. In 1994-95 the area and production rose to 83,000 hectares and 7,000 tons respectively. Currently, the area and production are at 71, 000 hectares and 13,000 tons, respectively.

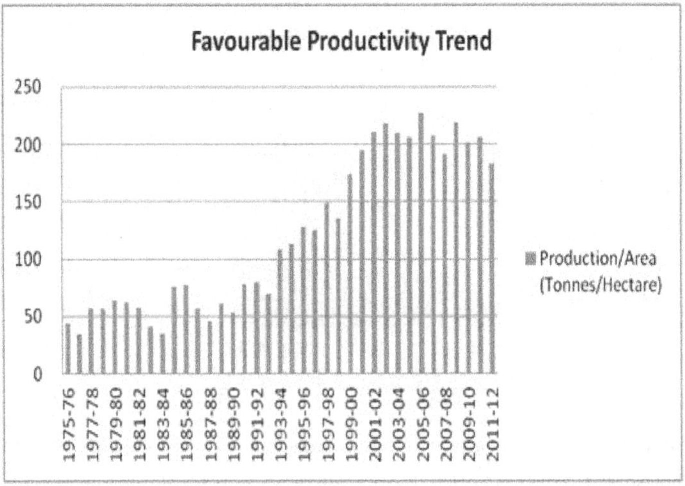

This was partly, as a result of the spread of a new variant of the indigenous Indian Cardamom. The variant is popularly called Najjani. This is a classic example of scientific interventions that saved the day.

The demand for cardamom will continue to rise, to meet the demands of a growing population and number of proven health and functional benefits with research. The question that the industry then faces is to make sufficient quantity available at the right quality. We should not let us concern constrain growth.

A steady supply of food safe spices of the desired quality would be a key factor of growth and we have to invest in technologies to ensure this. Five years back nobody wanted to put any money to integrate backwards–to make a food safe supply chain–today the need in the industry is well felt. Companies in Kerala engage in what they call IPM (Integrated Pest Management) programs to tightly control agricultural inputs, farm practices and thereby producing the quality they desire. There is a buy back arrangement entered into, and often the farmer obtains a premium.

For those things that cannot be accomplished by an individual company, associations such as the World Spice Organisation come into picture. WSO, has been working with farmers, scientists, processors, regulators and consumers, on projects that target different challenges: Awareness programs, joint discussions among different stakeholders to create a '*coalition of the willing*' to work towards a sustainable future for cardamom.

CHAPTER-III

CULTIVATION PRACTICES AND ISSUES

CARDAMOM (Elettaria cardamomum)

Characteristics of Plant

The habitat of small cardamom is the evergreen forests of Western Ghats. It is grown in areas where the annual rainfall ranges from 1500-4000 mm with a temperature range of 10-35 ºC and an altitude of 600-1200 m above MSL. Cardamom is generally grown in forest loam soils rich in available phosphorus and potassium. The crop is raised mainly on well drained, deep, good textured soils rich in humus.

Nursery

Ripe capsules of the desired cultivar are collected from high yielding plants during September-October. The seeds are extracted by gently pressing the capsules. In order to increase the germination percentage, seeds can be treated with concentrated sulfuric acid or nitric acid for not more than two minutes. The extracted seeds are washed in cold water four times to remove the mucilaginous coating. The washed seeds are drained and mixed with ash and allowed to dry in shade for 2 or 3 days. The seeds should be sown in the nursery within a fortnight. Sowing in September is the best for high germination. Sowing during southwest monsoon and winter should be avoided.

When it becomes necessary to store the seeds, it is advisable to store them in capsule form. It can be preserved in this form for one month, without deterioration of viability. Polythene lined gunny bags can be used for this.

In Kerala and Tamil Nadu, 18 month old seedlings are used for planting. The seeds are sown in primary nursery from where the young seedlings are transplanted to a secondary nursery and maintained for one year before planting in the main field.

Primary Nursery

The nursery site is selected in open, well-drained areas, near a source of water. The land is due to a depth of 30 cm, cleared of all stubbles and stones; and clods are broken. Beds of size 6 x 1 x 0.3 m are then prepared. Jungle soil is spread in a thin layer over the nursery bed. Seeds are shown on the bed in lines. For an area of 1 m2, 10 g of seed is required. Sixty grams of seeds will be required for a nursery bed of 6 m2. The seeds are covered with a very thin layer of fine soil. The nursery bed is mulched with dry grass. Potha grass (Grenetia stricta) commonly seen in high range areas is a suitable material for this purpose. Grass is spread to a thickness of about 2 cm. Paddy straw can also be used for mulching. After sowing, beds have to be watered every day in the morning and evening. The mulch should be removed on commencement of germination. The seedlings have to be protected by providing shade pandals. Regular watering, weeding and protection from pests and diseases are to be attended to. During June-July, seedlings from the primary nursery are transplanted to the secondary nursery.

Secondary Nursery

After preparing the site properly, form nursery beds of 6 x 1 x 0.3 m. Mixing of well decomposed cattle manure and wood ash with the top layer of the soil will help the seedlings to establish well and to grow vigorously. During June-July, the seedlings from the primary nursery are transplanted at a spacing of 25-30 cm. Shade pandals should be provided before transplanting. Overhead pandals or individual pandals for each bed may be erected. Mulching the bed with dry leaves will help to conserve soil moisture. Regular watering during dry months, weeding, application of fertilizers, control of pests and diseases and mulching are the essential operations for the maintenance of the secondary nursery. One month before uprooting, the pandal should be removed to encourage better tillering.

Polybag Nursery

Polybags can be used for raising secondary seedlings. For such nurseries, seeds are to be sown in beds in primary nurseries in September and transplanted to polybags in December-January. These seedlings would be ready for planting in June-July. In this case, nursery period could be reduced by 6 to 7 months.

Rhizome Multiplication

This may be taken up from the first week of March to the first fortnight of October. The site is selected in open, gently slopping and well-drained areas near a source of water. Trenches of 45 cm width, 45 cm depth and convenient length are taken across the slope or along the contour 1.8 m apart. They are filled with equal quantity of humus rich topsoil, sand and cattle manure. Uproot a part of the high yielding disease free mother clump identified in the plantation. Trim the roots and separate the suckers so that the minimum planting unit consists of one grown up tiller and a growing young shoot. Plant them at a spacing of 1.80 m x 0.60 m in filled up trenches. Provide sufficient mulch and stake each planting unit. Provide overhead pandal as in the case of seedling nursery and remove shading material with onset of monsoon rains. Provide irrigation once in a fortnight and adopt necessary plant protection measures. Apply fertilizers @ 100:50:200 kg/ha N:P2O5:K2O in six splits at an interval of two months. Apply neem cake @ 100-150 g/plant along with fertilizers. On an average, 20 to 30 suckers / initial planting unit can be produced within one year of planting. Care should be taken to identify and collect mother clumps only from areas totally free from 'katte' disease.

Soil Treatment in Nursery

It is recommended that the primary and secondary nursery soil may be drenched with formalin 2% solution and covered with polythene sheets for 3 days. Planting should be taken up only 15 days after treatment to avoid phytotoxicity.

Shade

Since inadequate as well as excessive levels of shade are harmful to the crop, regulation of shade is inevitable. There should be sufficient shade to protect cardamom plant during the hot season. By regulating the shade before the monsoon, more light becomes available to the plant during the rainy season. Red cedar or chandana-vempu (Toona ciliata) is an ideal shade tree. It sheds the leaves during rainy season and thus provides natural shade regulation. Some of the other shade trees are kurangatti (Acrocarpus fraxinifolius), vellakil (Dysoxylum malabaricum) and thelli (Canarium strictum).

Genetic Resources and Breeding

A total of 618 accessions are being maintained at NAGS and about 117 accessions were characterized for morphological and yield characters. FGB-13 and FGB-82 recorded maximum yield and more number of capsules per plant.

The accession, IC 547167 (Appangala 1 x NKE 19) with potential yield of 1393.12 kg ha-1 (three years after planting), with mosaic resistance and good quality characters has been recommended for release in Karnataka as new variety under the name Appangala 2 by XXV AICRPS meeting held at UBKVV, Pundibari, West Bengal in September 2014.

Standardizing the Parameters for Target Yield

In Green gold, recorded yield levels per plant basis was 0.7, 0.9 and 0.9 kg plant-1 for the targets 0.4, 0.6 and 0.8 kg plant-1 with a positive mean deviation of 72, 55 and 15%, respectively. Similarly, in Appangala 1 yield per plant has shown a positive mean deviation of 83, 76 and 14% for the fixed target levels.

Soil carbon pools under cropping systems

The total and particulate organic carbon (POC) and nitrogen pools were quantified under different spice based cropping systems and high density multiple cropping system. The POC and PON pools were higher in coffee + black pepper system (56.7 & 16.8 mg ha-1) with highest total organic C & N (TOC & TON) pools (90.1 & 33.4 mg ha-1). POC constituted 63% of TOC in this system. The non particulate carbon and nitrogen (NPOC & NPON) pools were higher under cardamom alone and coffee + black pepper + cardamom cropping system (67.3 and 58.3 mg ha-1) constituting 73-78% of TOC pools.

Among different management systems in black pepper, organic management has resulted in higher POC, NPOC and TOC pools as compared to integrated and conventional management systems. In HDMCS, black pepper basin has accumulated highest TOC, NPOC and POC pools (106.8, 71.6, 35.2 mg ha-1, respectively) and coconut and nutmeg systems had higher NPON and PON (7 & 0.8 mg ha-1) as compared to other component crops.

Quality evaluation of cardamom varieties

Different varieties *viz.*, Njallani, Pannikulangara-1, Pannikulangara-2, Thiruthali, Elarajan and Wonder cardamom collected from Idukki district were analyzed for essential oil profile. The essential oil content ranged between 5.8-7.4% on capsule weight basis. Pannikulangara-2 recorded highest essential oil content. Among the 21 components identified, the chief constituents of the oil, 1,8-cineole and α-terpinyl acetate varied between 18.1-32.7% and 36.9-48.5%, respectively. Concentration of pinene, sabinene, myrcene, α-terpineol, 4-terpineol, nerol, neryl acetate and nerolidol ranged from 1-5%.

Quality Evaluation Using E-nose

Hand-held electronic nose was modified with suitable sensor array for determining quality. Samples were analyzed using the modified hand-held electronic nose for essential oil content and could be graded into low (<4.0%), medium (4.0-6.0%) and high (>6%).

Management of Thrips (Sciothrips cardamomi)

Screening of natural products and newer insecticides for three years indicated that Spinosad, a natural product derived from Saccharopolyspora *spinosa* can be used for the effective management of thrips. The product can also be used in organic system. Standardization of spray schedule of promising insecticides is in progress.

Evaluation of Entomopathogenic Fungus

Field trials with the promising entomopathogenic fungus *Lecanicillium psalliotae* for the control of thrips were conducted at Kodagu, Wayanad and Idukki districts. The trials indicated that combined application of *L. psalliotae* as spray and basal application gave better control than other treatments at Wayanad.

Studies on Wolbachia

Studies on removal of the endosymbiont *Wolbachia* from thrips as a strategy for its management indicated that when the thrips were fed with tetracycline treated leaves, *Wolbachia* was completely eliminated from the insect system which was confirmed by molecular studies.

Root Grub (Basilepta fulvicorne)

Infectivity of entomopathogenic nematodes (EPNs) against root grub, *Basilepta fulvicorne* was tested *in vitro*. Among the test EPNs, *Heterorhabditis* sp. (IISR-EPN 01) and *O. gingeri* (IISR-EPN 07) were more pathogenic as they caused 100% mortality to the insect within 72h post exposure, followed by *Steinernema* sp. (IISR-EPN 03), *S. carpocapsae* (IISR-EPN 06) and *Oscheius* sp. (IISR-EPN 08). *Steinernema* sp. (IISR-EPN 02) and *Oscheius* sp. (IISR-EPN 04 and 05) took 120h to kill the test insect.

Occurrence of Perfect Stage of C. gloeosporioides

Surveys carried out in cardamom plantations revealed manifestation of different types of foliar symptoms *viz.*, spot, blight and shredding. The cultures isolated from these samples exhibited variations in colony morphology and colour.

Among the cultures, greyish white culture appeared puffy with a faster growth rate (14 mm day-1) and produced dark brown-black, globose perithecia, four weeks after incubation.

Differential Reaction of C. gloeosporioides Isolates on Varieties

Differential reaction of 20 *C. gloeosporioides* isolates was studied on cardamom varieties *viz.*, Appangala 1, IISR Vijetha and IISR Avinash. The isolates exhibited differential reaction as indicated by prominence and non-prominence of yellow halo and streak.

The area of lesions developed on young leaves varied between 4.91 - 40.82, 7.85 - 60.45 and 11.78 - 38.47 mm2 in IISR Avinash, IISR Vijetha and Appangala 1, respectively.

Evaluation of Microbes for Antagonistic Potential against rhizome and Root Rot Pathogens Under in Vitro Conditions

The endophytic fungal isolates from varieties IISR Vijetha, IISR Avinash and Appangala 1 were evaluated *in vitro* for antagonistic efficacy against

Fusarium oxysporum, *Rhizoctonia solani* and

Pythium vexans. Among the isolates tested, Va 4-2 (IISR Vijetha), Cb 4-1, Cb 6-2 (Appangala 1) and Aa 1-1 (IISR Avinash) were found promising against *F. oxysporum*. While, Cb 4-1, Cb 6-2 (Appangala 1) and Ab 6 (IISR Avinash) were effective against *P. vexans* and Cb 2 (Appangala 1) was inhibitory to *R. solani*.

Major Cultivate Areas

Malabar: Suitable for areas from 600 to 1200 m elevation

Mysore: Suitable for areas from 900 to 1200 m elevation

Vazhukka: Suitable for areas from 900 to 1200 m elevation

Seeds and Cultivation

Propagation

Cardamom can be propagated vegetatively and by seedlings. For vegetative propagation, rhizomes with not less than three shoots are used. Plants propagated vegetatively come to bearing one year earlier than the seedling-propagated plants. But this method has the disadvantage of spreading the `katte' disease, which is of viral origin. This disease is not transmitted through seeds. Hence in areas where the disease is widespread, it would be safer to use seedlings for propagation.

Main Field Planting

Cardamom plantation is raised in forests under the shade of tall trees. For raising a new cardamom plantation, the undergrowth of bushes is cleared. When open areas like marshy valleys and grasslands are selected for raising new plantation, shade trees have to be raised before planting cardamom seedlings. The quick growing shade trees like dadap (Erythrina lithosperma) is generally used for this purpose. Cuttings of this tree are used for planting. But this tree is a host of root knot nematode, which infests cardamom. Other quick growing trees like Albizia can also be used. Useful trees like jack and eucalyptus can be used along with red cedar, wild nutmeg, kurangatti etc.

Harvesting

Cardamom plants normally start bearing capsules from the third year of planting. Picking is carried out at an interval of 30 days. After harvest, cardamom capsules are processed.

Cardamom capsules with green colour fetch a premium price in foreign countries. Hence emphasis has to be given on the preservation of green colour during curing and subsequent storage. Capsule should be processed within 24-36 hours after harvest to prevent deterioration. By curing, the moisture of green cardamom is reduced to 8-12 per cent at an optimum temperature so as to retain its green colour to the maximum extent.

Harvesting season in Kerala is October-February and the peak period of harvest is September-November.

Sun-drying

Capsules are dried directly under sunlight for five to six days or more. Frequent turning is done. This method can result in surface blemishes and may not give an attractive green colour. This method is practiced if the cultivar yields fruits that turn yellow before they are ready for picking and where facilities for green curing are not available.

Artificial Drying

Processing of capsules is done in specially built curing houses. The harvested capsules are washed in water to remove dust and soil particles. Then they are spread on wire net trays in curing chamber. Burning firewood in the iron kiln produces heat required for drying. The heat thus produced is passed through pipes made of galvanized iron sheets. The process of drying takes about 18-24 hours, depending on the ambient temperature. The capsules are spread thinly in the wire net trays and stirred frequently to ensure uniform drying. They are initially heated at 50 ºC for the first 4 hours and heat is then reduced to 45 ºC by opening ventilators and operating exhaust fans till the capsules are properly dried. Finally the temperature is raised to 60 ºC for an hour.

The dried capsules are rubbed on wire mesh to remove the stalk and dried portion of flower from the capsules and then graded according to size by passing through sieves of sizes of 7, 6.5, 6 mm etc. The graded produce is stored in polythene lined gunny bags to retain the green colour during storage and also to avoid exposure to moisture. A relatively new innovation in the curing procedure is blanching by soaking the fruits in 2.0 per cent washing soda for 10 minutes prior to drying. This inhibits colour loss during drying operation and extends colour retention during subsequent storage from three months to ten months.

Crop Seasonality

Cardamom plants normally start bearing fruit two years after planting. Throughout the cropping season of cardamom, from August to March, there are approximately 6 pickings at an interval of 45 days. The peak harvest period is during September- November. Ripe capsules are harvested inorder to get maximize the green colouring of the pod during curing. Cardamom output is greatly influenced by climatic conditions. The plant requires a continuous spell of rain, intermingled with periods of good sunshine. The plant is very susceptible to attack by pests and diseases.

Besides, the cardamom growing tracts in the country have been facing severe ecological degradation over the past two decades due to diminishing forest cover, leaving the regions open to devastation by floods and droughts. Cardamom requires tropical forest conditions for growth and, a lack of such areas makes very few states in the country suitable for the plant.

Labour Shortages

The plantation sector has been facing difficulty in obtaining and retaining workers in the sector. Despite the wages being competitive the supply of workers have remained /declined in the plantation sector. This has been one of the banes of the sector. This note intends to throw light on the issue of labour shortage in the plantation sector by analyzing the employment and wage conditions in the sector in specific and the related economy.

Employment

Plantation is a part of agriculture and it directly employs more than 2 million workers in the country. As per the study on Socio-economic Conditions of Women Workers in Plantation Industry, 2008-09 about one million workers are involved in the plantation sectors of coffee, Rubber and Tea in 2005-06 (Table 11.1). Of this nearly eight lakhs are engaged in the Tea sector while one lakh workers are in the Coffee and about 60000 workers in the rubber plantations. In general unlike in most other sectors women labour (54 percent of workface) is more in the plantation sector compared to male workers. However in case of Rubber plantations the workforce is male dominated with nearly 58 percent being male workers The average daily employment in the plantations shows that on an average the employment per day had declined continuously during the decade 1990-2003 (Table 11.2). The average employment declined from one million workers per day in 1990 to 6.7 lakh workers in 2003. However this declining trend is most visible in case of Tea, and coffee. While for Rubber during this period there is actually a rise in employment from 24 thousand workers in 1990 to 30 thousand workers in 2000.Along with this trend in decline in employment the share of women workers had been increasing in all plantations. It increased from 51.4 % in 1990 to 53.2 % in 2003. The decline in overall employment has been attributed to the poor performance of coffee and Tea production during these periods. However, the rising proportion of women workers in the rubber sector, traditionally a male dominated sector, along with increasing employment in the sector is indicative of the bullish growth and the employment demand in this sector.

Average Daily Employment in the Plantations that are Covered Under the Plantation Labour Act

		1990	**1994**	**1998**	**1999**	**2000**	**2001**	**2002**	**2003**
Coffee	employment	29341	22527	28333	26137	29249	31237	22769	21872
	%share of women	53.82	60.31	55.88	50.27	47.79	55.35	57.84	58.39
Rubber	employment	24583	35272	22092	18831	26556	28164	29329	30789
	%share of women	38.08	41.36	33.61	34.23	40.37	41.26	42.53	42.17
Tea	employment	1024979	1687128	894932	852675	903024	322004	665554	615195
	%share of women	51.72	35.43	48.89	49.79	50.05	53.64	53.48	53.61
Other	employment	5677	5077	5649	4488	6784	7557	4223	5524
Plantations	%share of women	49.18	56.25	46.61	51.31	52.89	48.95	54.84	49.96
Total	employment	1084580	1750004	951006	902131	965613	388962	721927	673380
	%share of women	51.45	35.93	48.73	49.49	49.73	52.79	53.18	53.21

Source: Statistical profile on Women labour 2004

CHAPTER-IV

PRODUCTION &MARKETING

Global Scenario

The world's cardamom production for the year 2013-14 is projected to be at 35000- 40,000 MT. At present, the major producer is Guatemala, is expected for produce around 25,000+ MT. India is the second largest producer with an average production of 12,000 to 12,500 MT.

Other major producing countries are Sri Lanka and Tanzania. Despite its numerous applications in the cuisines of Sri Lanka, India and Iran, 60% of the world production is exported to the Arab (South West Asia, North Africa) countries, where the larger part is used to prepare coffee.

Indian Production Scenario

In India, two types of cardamom are grown; small and large. The production of small cardamom is confined to the southern states while the large variety is grown mainly in the north-eastern states. Among the major small cardamom growing states, Kerala is the largest part of the pie at 76% of the total production followed by Karnataka (16%) and Tamil Nadu (8%). Moreover, weather conditions in Kerala are very perfect for cardamom, which makes it the largest cardamom growing state in India.

Domestic Consumption Pattern

At present, India is the second largest consumer of small cardamom in the world after Saudi Arabia.

India's consumption is estimated to be between 11,000-12,000 MT per annum in 2013-14. Industrial consumption of cardamom, particularly by the pharmaceutical or ayurvedic and cosmetics industries is the highest and accounts for over 45% of the total consumption. A clear regional disparity is seen in cardamom's consumption pattern. Though the urban areas contribute to a large share of its consumption, the rural markets have been gaining importance in recent years and growing at a higher rate than the urban market.

Household uses of this spice are very limited, mainly due to its high price. However, seasonal buying has increased.

Cardamom Balance Sheet - India

Year	**Supply(In Tonnes)**		**Demand(In Tonnes)**		
	Beginning Stocks	Production	Consumption	Export	Ending stocks
2009-10	430	10500	8000	1975	50
2010-11	50	12500	10500	1175	875
2011-12	875	20000	12500	4650	3725
2012-13	4000	14500	14000	2372	2128
2013-14	2128	12500	10420	2080	2128

Export Data-2013-14

Export of Cardamom from India

India is the second largest exporter of cardamom after Guatemala. Indian small cardamom is always preferred in the global market due to its aroma and premium quality. India exports roughly 15% to 20% of its total cardamom production. Saudi Arabia is the single largest importer of cardamom, distantly followed by Kuwait. However, export during 2004-05 to 2008-09 remained lower due to higher domestic consumption and production of less than 10,000 MT. Nonetheless, in the year 2009- 10, exports from India shot up to 1975 MT due to a drastic fall in production amid unfavorable weather conditions. In 2010-11, export fell to 1175 MT but, during 2011-12 due to an increase in production, exports have become easier. In 2011-12, India exported around 4,650 MT of small cardamom, which is considered the record high of the past few years. However, during the year 2013- 14, 2,080 MT of cardamom valued at Rs 16,429.38 lakh were exported, compared to the 2372 MT in 2012-13 according to the data released by the spices board of India.

System of Innovation in Cardamom Marketing and Spaces of Exclusion

Cardamom is a plantation crop grown historically in the evergreen forests mainly for export and its price depends on the quality of the dried capsule. The quality of the product in turn depends on the colour, aroma, and boldness of the capsule. All these depends on the plant variety, timing of harvest (which is a highly skilled and women labour intensive) curing method adopted (sun drying or in curing houses) and timely implementation of cultural operations especially spraying of pesticides to protect the capsules from a number of insects and pests. To the extent that there is significant variation across and within holdings in terms of harvesting, curing and other cultural operations along with plant variety there is significant quality variation across and within output offered for sale by a given grower and across different growers. Given such quality variation an auction system has been followed historically for the marketing of cardamom[4]. Until the introduction of e-auction in 2009 in Bodinaikanur (Tamil Nadu) and 2010 in Puttadi (Kerala) auction has been in the traditional outcry method wherein the bidders competed face to face for the lots offered for sale. In what follows, we shall begin with a discussion of the traditional auction system followed by the e auction.

The institutional innovations in the marketing of commodities like cardamom at the primary level take place in the form of state regulation. A market is said to be regulated when the government establishes it under some enactment, and frames rules and regulations to conduct business there in (Harris 1981). The nature of regulation is found to be varying from crop to another. In case of cardamom, the regulation takes the form of restricting the entry of different functional categories in the like of the sellers, auctioneers, dealers, and exporters into the auction system as per the Cardamom (Licensing and Marketing) Rules of 1977. The declared objective of such a regulation has been to ensure a fair price and timely payment of sale proceeds (Nair et al 1989, Narayana 1994, Joseph 1985).

The marketing system for cardamom market has been characterised by large number of sellers (both large and small) a small number of exporters – many of them are dealers and few of them, in addition to being dealers, are growers as well – a large number of dealers and a few auctioneers. Auctioneers are one of the major actors in the marketing of cardamom who bring together the buyers and sellers of the product. During 1980s when the auctions were conducted in the conventional outcry manner and held in the premises of the auctioneer, there were 14 auctioneers in India spread over the cardamom growing States, viz. 5 in Kerala, 2 in Tamil Nadu and 7 in Karnataka. Vandanmettu in Kerala, Pattiveeranpatty in Tamil Nadu and Sakalespur in Karnataka were the major auction centres in India.

The auctioneer brings together the producers holding registration certificate and the dealers (also referred to as 'bidders') holding license issued by the Spices Board to deal in cardamom. As per the cardamom rules, the dealers shall not purchase cardamom from an estate owner who has not registered his estate or from an auctioneer who has not been licensed by the Spices Board (formerly Cardamom Board). Further, no dealer shall solicit or accept any amount in cash or in kind from cardamom producers or auctioneers whether by way of discount or commission. A dealer participating in auction shall pay full value for the entire quantity of cardamom in the lot purchased by him.

The cardamom exporters are another important link in the marketing system of cardamom. According to the Cardamom (Licensing and Marketing) Rules of 1977, the exporters shall not procure cardamom directly from a grower or from an auctioneer unless he is in possession of a license as dealer. The number of exporters varies from year to year as the license is issued for a year after which it should be renewed to carry on business.

Cardamom market was characterised by high level of concentration at all levels. During 1980s, according to an earlier study, though there were 5 auctioneers in Kerala, one of them (CMC Vandamettu) accounted for more than 70 percent of the product sold through auctions. Apparently, the availability of large supply attracted large number of bidders and the consequent increased degree of competition pushes the price up. The realisation of higher price in the particular centre induces more sellers to register their product with that Centre.

It was further observed that during 1971-85 only two auctioneers remained permanently in business and together they accounted for nearly 90 percent of the total sales in Kerala. It should also be noted that all these auction centres were located in the cardamom growing areas of Idukki District covering an area of 45,172 hectares in 18,585 holdings. The only exception was the auction centre at Cochin. In Waynad there was no auction centre though there are 270 holdings with an area of 4,247 hectares - a case of active exclusion according to the taxonomy of exclusion suggested by Amartya Sen. It is also to be noted that for historical reasons, larger number of cardamom growers are not having title deeds and that they are not entitled to register their product with an auction centre - yet another case of active exclusion.

In 1983-84 there were 528 licensed dealers in India. It was found that while Kerala accounted for 69 percent of the total production, only 32 percent of the total dealers belonged to Kerala. There was a clear domination of Tamil Nadu in cardamom trade. That is, Tamil Nadu, with 6 percent of the total production accounted for 43 percent of the dealers. Historically, the activities of the traders were confined mainly to two major markets of Tamil Nadu, viz. Bodinaikannur and Viruthnagar. The situation remains more or less the same at present.

The dealers buy cardamom mainly at auctions and directly from growers. Even though all the traders are licensed to deal in cardamom marketing at every stage, all of them are not equally involved in trade. Only a few of them, about 125, used to deal in auction business. This could be because of the need for huge capital and risk associated with auction. Therefore, many of these dealers preferred to buy directly from growers. The dealers perform an important function of grading the product. They grade and pack it according to the demand from different consumers within the country and outside.

With respect to concentration of market power, it was found that on an average 29 dealers bought cardamom from the major action centre (that accounted for over 60 per cent of the output sold in Kerala) during 1979-80 to 1983-84. Out of this, the share of top 5 dealers was found to be 58 percent in 1979-80 and it increased to 65 percent in 1983-84. Similarly, the share of top 2 dealers also increased during the same period. It was 34 percent in 1979-80 and it increased to 39 percent in 1983-84. This may be partly accounted for by the decline in the number of bidders participating in the auction. All the top 5 bidders are found to be exporters. Of these five, one dealer is found to be dominating in all the years (Joseph 1985).

At a time when cardamom was an export oriented crop[5] the price realised in the internal market depended to a great extent on the market strategy adopted by the exporters. This was more so, as the export trade was controlled by a few exporters. Export in 1976-77 was 893[6] M.T and it increased to 2876 M.T in 1978-79 and showed a marked decline in 1982-83. With increase in the quantity exported the total number of exporters increased and as quantity exported declined the number of exporters also decreased. The interesting point to note is that while the average share of an exporter was nearly one percent, the top 10 exporters accounted for more than 60 percent of the total and some of these leading exporters are found to be the dominant buyers in certain important auction centres. Moreover, their share in the total export increased considerably during the period under consideration. Even though the export was subjected to yearly fluctuations, the share of top 5 exporters was reasonably stable at 50 percent in all the years under consideration (Joseph 1985).

Thus the foregoing discussion reveals that prior to the introduction of e-auctions, the institutional innovations has had the effect of concentration of market power in the hands of a few at all levels. A few auctioneers accounted for bulk of the quantity sold through auctions and a few dealers accounted for bulk of the quantity bought through auctions. At the export level also a handful of exporters controlled most of the quantity exported.

Functioning of the Market

Having examined the marketing system, let us proceed to examine the functioning of the market which was governed by the institutional innovations in the form of marketing rules brought about by the erstwhile cardamom board. The growers register their produce with the auction centre, sufficiently before the auction, and may get an advance from the auctioneer which will be deducted from the sale proceeds. The auctioneer maintains a warehouse where the produce is kept till the time of auction. If the product is withdrawn from the auction, (the seller can withdraw the product if the price quoted is unsatisfactory) the seller is liable to pay the warehouse charges at the rate of 15 paise per kg per day.

At the time of registering the produce for auction, the auctioneer mixes the produce thoroughly and takes a sample from each lot. The Cardamom (Licencing and Marketing) rules (1977) describes the procedure of taking the sample as follows:

"The auctioneer shall draw 500 grammes out of each lot of cardamom offered for sale as sample. Expose 350 grammes out of it for bidders to examine at the place of auction four hours before the commencement of the auction and circulate a list indicating quantity of each lot, weight in grammes per litre[7] and reserve price[8]. The 100 grammes out of the sample shall be given to the highest bidder and the balance of 50 grammes shall be kept by the auctioneer in a sealed polythene bag for a period of seven days for verification in case of any dispute. On expiry of the said period or on settlement of the dispute, as the case may be, the owner of the cardamom shall be entitled to receive back the cardamom from the auctioneer."

But the actual procedure of taking sample and auctioning was stated to be different from what is stipulated by the rules. The auctioneers in Kerala take a sample of 750 grammes of which 500 grammes is paid at the rate at which the rest of the quantity is sold. Out of the 250 gramms, 100 grammes is given to the highest bidder and the rest 150 grammes is appropriated by the auctioneer in addition to the one per cent commission. Further, the auction report showed that 8 percent of the sales tax is taken from the seller as handling charges. Thus from each lot, whatever may be its size, 250 grammes is lost by the seller in addition to 8 percent of the sales tax as handling charges. The practice of deducting fixed quality from each lot as sample irrespective of the lot size, make the effective price (the price that the sellers get per kg after all deductions) much less than the actual price (the price quoted in auction). This discrepancy between actual price and effective price reduces as the lot size increases. Thus the registered sellers with a lot size below a certain minimum quantity may not find it profitable to sell their produce through the auction centres because of the institutionally set rules; a typical case of active exclusion that Amartya Sen (2000) described.

Under the conventional auctions, it used to take place in the premises of the auctioneers and procedure used to be as follows. The dealers (bidders) will be seated around a round table and a small tray will be kept in front of each of them, and out of the 750 grammes of cardamom taken as sample from each lot, 500 grammes will be spread in all the trays for examining the lot. Then the intending buyers bid upwards for individual lots and the highest bidder will be given the lot. The process of bidding starts with the announcement of the lot number and lot size. It is a quick process and each lot takes hardly a minute to get auctioned. The seller has the freedom to withdraw her product from auction before the bidding is over. Once the price is determined through bidding, the seller loses his right over the produce.

Once the bidding is completed, a slip showing the weight of the lot and the agreed price is given to the seller by the auctioneer, not the value of product sold. The slip also shows all the deductions and charges to be paid by the seller. The payment is made later, usually after 20 to 45 days as the case may be. There is, thus, an in-built credit facility extended by the growers which in turn is guaranteed by the auctioneers. The possible explanation for this delay may be the huge capital requirement on the part of the dealers because of the high unit value of the product. Therefore, the growers will have to wait for the product to be exported or to be dispatched to the consuming centres. Some of the growers were of the view that the delay in payment has been perpetuated both by the dealers and the auctioneers. The auctioneer could refrain from paying to the sellers for the single reason that a few of the dealers have not made the payment even though all the others have paid. The delay in payment, therefore, keeps away those growers without the holding power (presumably the small holders) from the auction system a typical case of illusive exclusion.

A perusal of the auction procedure described above reveals that there are different types of exclusion inherent in the system resulting from the institutional innovations. There were a large number of growers who were not having the registration certificate and hence confronted with active exclusion and active exclusion also take place on account of the institutionally set practice of taking fixed sample from each lot regardless of the quantity offered for sale. From our discussion with the growers we came to know that they are forced to sell their produce at the auction centres, even at an unsatisfactory price, because of the inability to repay the advance already received by them, a typical case of illusive inclusion. If they withdraw the product they will have to pay the go down charges at the rate already mentioned. Thus the practice of giving advance at the time of registering the product for auction seems to have the effect of subordinated inclusion.

It is pertinent to note that, the system of taking sample which prevailed in the auction centres in 1940s was on the basis of quantity sold. For example, in Bodinaikannur, a sample of 2 lbs was taken from lots up to 10 thulams (1 thulam=10 kg) and 3 lbs from the lots above 10 thulams and the payment was made at the time of sale. Therefore, the effective price realised was proportional to the price quoted the same for all sellers. In other auction centres rather than taking any sample, a service fee of 8 annas per thulam was charged. Needless to say, since the service charge was fixed per unit quantity, irrespective of the size of the lot, the effective price received by the sellers was proportional to the auction price (Joseph 1985).

Price Variation across Different Lots under Conventional Auctions

Having examined the spaces of exclusion induced by the institutional innovations let us now examine the process of price formation in auctions and price variation across different size of lots sold through auctions to see if there are additional spaces exclusion being created.

To begin with, let us be clear about the market power of different actors involved; from the sellers side there is the market power of the growers of cardamom, both the large and small, who could withdraw the product from auction if the price quoted in the auction is found to be unsatisfactory. The auctioneers, similar to the sellers, are also interested in obtaining higher prices for the following reasons. First and foremost, since the auctioneers were also the growers of cardamom, higher prices would lead to increase in their sale proceeds. Secondly, higher the price realised in auction, higher the commission accruing to them and thirdly, higher price quotation in a certain auction centre would attract more sellers to that centre thereby increasing the total revenue accruing to the auctioneers by way of taking sample.

Coming to the market forces operating on the buyers' side it needs to be noted that exporters, as a category, consist of both large scale exporters and small scale exporters and the leading exporters are found to be the major buyers in the auction centres. Since the profit of the dealers and exporters depends on the margin between the export price and auction price, the dealers and exporters would try to depress the auction price with a view to increase their margin. This is because the export price has been exogenously determined and therefore the exporters may not be able to manipulate the export price. Hence, in the bidding process in the auction centres, the exporters quote the price in accordance with their price agreements with the importers to other countries.

Under these conditions one could postulate the process of price formation in auction centres in the following way: In the bidding process the bidders (the exporters and dealers) would endeavour to reduce the price, given the fact that they have already made commitment to their counterparts at a certain price, with a view to add to their margin. At the same time, the growers and the auctioneers would be interested in obtaining the maximum possible price. If the price quoted is unsatisfactory the sellers would withdraw the product from auction. The relevant question is, which category of sellers will be able to withdraw the product? Obviously, only those with adequate holding power and good storage facility could withdraw the product. On the other hand, those sellers without adequate holding power and who have already received an advance from the auctioneer will be forced to sell the product at the price quoted in the auction. Thus, there is the possibility of price discrimination across different lots sold through auctions wherein the smaller lots (smaller growers) realising a lower as compared to the larger lots.

In this context, the pertinent question is to what extent the bidders could discriminate across different lots? This depends on the supply and demand conditions in the market. Given the export demand, if the domestic production is higher than what the export demand warrants (excess supply) the price quoted in the auction centres will be low and the price discrimination tends to be higher. On the contrary, in a situation of low production (excess demand) the price quoted in the auction centres would be relatively high and the price discrimination across different lots may be low With a view to empirically verifying the above hypotheses we have classified the lots sold through auctions during the two peak season months and two lean season months into different size classes and estimated the average price obtained by each size class of lots during 1979-80 to 1983-84 (see Table 1). The table throws light in to different interesting aspects of price variation.

Firstly, there existed a positive relationship between lot size and average price. That is, as we move upwards from the smaller size class, the average price obtained also rises. Secondly, the difference between smaller lots and larger lots is higher during the peak season and lower during the lean season. Thirdly, as we move from a year with high production to a year with low production, the extent of price variation across different lots gets reduced and finally unlike any other agricultural commodity, the peak price has been associated with peak season.

With a view to explain the price variation across different lot sizes, we have identified the quality of the lot and size of the lot as explanatory variables. The quality of the lot, in turn, depends on colour, aroma and boldness of the capsules. Since these variables are unquantifiable, we took the weight per litre of cardamom, as a proxy, to represent the quality of the lot. We estimated the following model by taking the price per kg as the dependent variable and the weight per litre and lot size as the independent variables.

$$P = a + \beta_1 Q_T + \beta_2 Q_L + e$$

Where

P = Price per kg of cardamom

Q_T = lot size

Q_L = Weight per litre (quality)

e = error term

The estimated price equations are given in Table 2. The result of the estimated model reveals the different role of quantity and quality on price variation as we move from peak season to lean season. The year 1983-84 for which the model has been estimated, was characterised by excess demand (on account of poor harvest) the price variation in the peak season could be explained by quality where as during the lean season it was mainly due to lot size. Thus the foregoing analysis leads to the conclusion that while the quality explains the price variation across different lots during the peak season, it is the changes in the lot size which explain the price variation during the lean season. This is indicative of the prevalence of subordinated or unequal inclusion that we have articulated in section 2 of this paper.

Price Variation Under E-auction

In a context wherein the traditional auction system has not been successful in ensuring remunerative prices for the small holders the Spices Board introduced e-auction in place of convention auctions with a view to make the marketing system more efficient and ensure remunerative prices to all growers by harnessing information technology Under the e-auctions, major change is in the auctioning system and that there is hardly any change in the rules governing auctions or the way in which samples are taken or in the time of payment. Unlike in the earlier system, auctions do not take place in the premises of the auctioneer, instead in a state of the art auction centre established by the Spices Board with over 60 computer terminals networked to a server and a few large display boards that display the details regarding the lots being auctioned (like reserve price, lot number, quantity, number of bags current highest bid). In the new system, licensed dealers are provided with a user id and password. The dealers have to login to the system to participate in an auction. In each week a particular day is assigned for each auctioneer for auctioning the product registered with them. The auction starts with the reserve price and a bid by a dealer is made with key depressions (each depression indicates an addition of fixed amount). As in the previous system the grower has the freedom to withdraw the product if the price is not satisfactory. Identity of bidders is protected during the auction process. Highest bidder's name is displayed only on the Auction Masters' terminal.

It should be noted that by the time e auction was introduced in 2010 in Kerala, significant changes had taken place in terms of total production, productivity and export intensity. There has been a significant increase in the production of cardamom from about 4400 MT in mid 1980s to over 10,000 MT in 2009-10. Similarly, productivity increased from about 55 kg per hectare in 1980-81 to 269 Kg per hectare in 2009-10. More importantly, share of export in production (export intensity) steadily declined from over 60 per cent in the mid 1980s to less than 10 per cent since 1990s.

There have also been changes in the market structure. It is evident from Table 4 which is based on the e-auction data obtained for the year 2009-11 from two auction centres (Puttadi in Kerala and Bodinaikanur in Tamil Nadu) that accounted for nearly 90 per cent of the total output sold, bulk of the output is handled by five of the auctioneers. Thus the domination of a single auctioneer prevailed in the mid 1980s is replaced today by an oligopolistic market structure (see Table 4). When it comes to market concentration at the level of dealers there appears to have been significant decline. While a few traders accounted for bulk of the quantity sold during the mid 1980s, today the share of top five dealers did not exceed more than 20 percent in none of the years and the share of largest dealer was only around 5 per cent in all the three years.

Trend in Production, Productivity, Export and Export Intensity of Cardamom

Year	Production (MT)	Export (MT)	Export intensity (%)	Productivity Kg/hectare
1975-76	3000	1941	64.7	44
1980-81	4400	2345	53.3	55
1985-86	4700	3272	69.6	56
1990-91	4750	400	8.42	79
1995-96	7900	527	6.67	122
2000-01	10480	1545	14.74	247
2005-06	12540	650	5.18	318
2009-10	10075	1975	19.6	269

Source: Estimates based on the data obtained from the different Annual Reports of Spices Board, Cochin.

With a view to understand the price variation across different lots under the e-auctions we have analysed the data on all the lots sold through the two e-auction centres during 2010. Average price obtained by lots belonging to different class of lots for both the seasons .It is evident that with the introduction of e-auctions there has been significant reduction in the price variation across lots belonging to different size classes. In contrast to the significantly higher price realisation recorded for the larger lots as compared to the smaller lots under the conventional auctions, we find that for the whole year and the lean season the lots in the smallest size class received higher price than that was received by the largest lots. But during the peak season price received by the larger lots is found to be higher than the smaller lots. What is interesting to note is the significant reduction in the extent of price variation between larger and smaller lots and that the extent of price variation at present is marginal when compared to what prevailed during 1980s. Also, in contrast to what was observed during the period wherein the product was export oriented, the peak price is not associated with the peak season. On the whole the empirical evidence indicates that the introduction of e-auctions appear to have had the effect of reducing significantly the price wedge between different lots of cardamom sold through auctions and hence the subordinated exclusion experienced under the earlier system With a view to analysing the bearing of different factors in price formation under e-auction we have estimated a model similar to the one estimated in the case of conventional auction. From the e-auction data provided by the auction centre, in addition to lot size and quality (measured in terms of weight per liter of cardamom), we also had data on the number of iterations and total quantity placed for auction. The number of iterations indicate the number of biding that took place for each lot and therefore it indicates the extent of competition for each lot sold. It is evident from the estimated co-efficient that all the four variables considered are found statistically significant in determining the price of cardamom sold. The value of the estimated coefficient, its sign and the level of significance have shown similar pattern during the peak season, and for the whole year. The results therefore tends to suggest that unlike in the conventional auction, during both peak season and lean season and for the whole year, prices are determined on the basis of the considerations like quality of the product, quantity available for sale and the extent of competition. In all the seasons we also find a positive relationship between lot size and price. Also as expected we observed a negative relationship with total quantity put for auction. The positive relationship that is evident in 2009 has to been against the fact that it had been a year with 10 per cent decline in domestic production. However, as we have already seen the price variation between smaller lots and larger lots is significantly lower as compared to the conventional auctions where it was found significantly higher. Moreover, the estimated correlation between lot size and quality is also found to be positive indicating that the larger lots that received higher price are also of better quality. On the whole, the analysis tends to suggests that the price discrimination and the subordinated inclusion that prevailed under the conventional auctions cease to exist under the e-auction.

Perspective and Recommendation by Horticulture

Background Facts: The purpose of regulation of agricultural markets was to protect farmers from the exploitation of intermediaries and traders and also to ensure better prices and timely payment for his produce. With this in view, only State Governments were permitted under provisions of APMC Act to set up markets and it was made mandatory to carry out sales-purchase of agriculture produce within the area of operation of regulated markets only inside the markets premise. Management of APMC's does not carry out auction-functions directly instead, gives licenses to brokers/commission agents known as "Adhatiyas" for carrying out auction of produce brought to the market. "Adhatiyas" are responsible to pay the price of the produce to the seller and has to recover purchase price from the buyers. As most of the trade of perishable is on credit, the "Adhatiyas" tend to manipulate sales of farm produce to traders known to them and who have better track record in making payments. This more often than not, results into lack of transparency in auction of agriculture produce, absence of real-time, truthful market information regarding price and arrivals of produce etc. The licensed brokers in the regulated markets have developed monopoly and political clout who do not allow easy entry of new persons, stifling the very spirit of competitive functioning.

Apart from the issue of lack of transparency in auction system, even the business process adopted by the market committees is generally insensitive to perishable nature of fresh horticulture produce. As per ruling business process adopted by regulated markets, on one hand, farmers have to bring their produce to the market yard where they do not get any value-added services and on the other, buyers like exporters,processors and retail chain operators who may not get desired quality and quantity of produce for their business; cannot buy the produce at the farm or at the processing plant or warehouse; the produce is required to be transported from the farm to the market yard and then only it can be purchased. In this process it becomes difficult to maintain traceability of the produce which is becoming one of the important requirements in export-markets.

On front of development of infrastructure, it may be observed that very few agriculture produce markets are specially designed for dealing with fresh horticulture produce that are highly perishable in nature and require proper handling, efficient sales, proper storage and further transport. The regulated markets were expected to plough back their revenue earnings into development of crop specific infrastructure; however, to the contrary, these markets have, over a period of time, acquired the status of restrictive and monopolistic markets and have not developed crop specific infrastructure and business process. Structural Limitations in Effecting Market Sector Reforms-Two main existing legislative instruments with the central and state governments, viz, the Agricultural Produce Marketing(Regulation) Act and Essential Commodities Act,1955 respectively are used to monitor the activities of market functionaries; however, provisions of the two Acts do not deal with the identified gap in agriculture /horticulture marketing system. Therefore, need for putting in place a new legislative frame-work providing for agriculture/horticulture produce marketing was felt. As per arrangement under the Constitution of India, agricultural marketing is a state subject; therefore, market reforms for agriculture/horticulture produce which has been a long felt need, can be brought about only by respective State Governments. Therefore, the Central Government had decided to effect market sector reforms by putting in place a Model Agriculture Produce Marketing (Development & Regulation) Act and has been encouraging States to adopt the same by various means such as making the adoption of Model Act as a pre-condition for States being eligible for getting benefits of Central Schemes relating to infrastructure development for agriculture produce markets. This has operated as incentives for State Government to adopt Model APMC Act which in turn had been expected to trigger the process of market sector reforms and promote private sector investment in setting up modern agriculture markets with proper PHM infrastructures, transparent auction system etc which may reduce "inefficiency" of value chain management.

Salient Features of Model APMC Act

- The Preamble of the Model Act sets out the objective to provide for development of efficient marketing system, promotion of agriprocessing and agricultural exports and to lay down procedures and systems for putting in place an effective infrastructure for the marketing of agricultural produce.
- Under the existing law, markets are setup at the initiative of State Governments alone. To the contrary, the Model Act provides that legal persons,growers and local authorities are permitted to apply for the establishment of new markets for agricultural produce in any area. Consequently, in a market area, more than one market can be established by private persons, farmers and consumers.
- It also prescribes that there will be no compulsion on the growers to sell their produce through existing markets administered by the Agricultural Produce Market Committee (APMC); however, agriculturist who does not bring his produce to the market area for sale will not be eligible for election to the APMC. Separate provision is made for notification of 'Special Markets' or 'Special Commodities Markets' in any market area for specified agricultural commodities to be operated in addition to existing markets.
- Even the APMC have been made specifically responsible for: ensuring complete transparency in pricing system and transactions taking place in market area; providing market-led extension services to farmers; ensuring payment for agricultural produce sold by farmers on the same day; promoting agricultural processing including

activities for value addition in agricultural produce; and publicizing data on arrivals and rates of agricultural produce brought into the market area for sale.

- A new Chapter on 'Contract Farming' has been added to provide for compulsory registration of allcontract farming sponsors, recording of contract farming agreements, resolution of disputes, if any, arising out of such agreement, exemption from levy of market fee on produce covered by contract farming agreements and to provide for indemnity to producers' title/possession over his land from any claim arising out of the agreement. Provision has also been made for direct sale of farm produce to contract farming sponsor from farmers field without the necessity of routing it through notified markets.
- In addition, accepting the principle of "State as One Market", provision has been made for imposition of single point levy of market fee on the sale of notified agricultural commodities in any market area and discretion provided to the State Government to fix graded levy of market fee on different types of sales.
- Under the provisions of Model Act, licensing of market functionaries is dispensed with and a time bound procedure for registration has been laid down; registration for market functionaries has been provided to operate in one or more than one market areas; commission agency in any transaction relating to notified agricultural produce involving an agriculturist is prohibited and there will be no deduction towards commission from the sale proceeds payable to agriculturist seller.
- Provision has also been made for the purchase of agricultural produce through private yards or directly from agriculturists in one or more than one market area.
- Provision is also made for the establishment of consumers'/farmers' market to facilitate direct sale of agricultural produce to consumers, and for resolving of disputes, if any, arising between private market/ consumer market and Market Committee.
- State Governments have been conferred power to exempt any agricultural produce brought for sale in market area, from payment of market fee.
- Market Committees have been permitted to use its funds among others to create facilities like grading, standardization and quality certification; to create infrastructure on its own or through public private partnership for post harvest handling of agricultural produce and development of modern marketing system.

Status of Adoption of Model APMC Act- In pursuance of above mentioned effort, the States/ U.Ts of Andhra Pradesh, Arunachal Pradesh, Assam, Chhattisgarh, Goa, Gujarat, Himachal Pradesh, Jharkhand, Karnataka, Madhya Pradesh, Maharashtra, Nagaland, Orissa, Rajasthan,

Sikkim and Tripura have notified the reforms in their APMC, Act in the key areas viz Direct Marketing; Contract Farming and Establishment of Markets in Private/Co-operative Sectors as suggested in the Model Act circulated by Department of Agriculture and Cooperation. The APMC Act implemented by State Government of Tamil Nadu already provides for the reforms as suggested in the model Act circulated by Department of Agriculture and Cooperation. There is no APMC Act in the States/UTs of Kerala, Manipur, Andaman & Nicobar Islands, Dadra & Nagar Haveli, Daman & Diu, and Lakshadweep. The Government of Bihar has repealed the APMC Act in the State. Only the State Governments/UT of Haryana, Punjab, NCT of Delhi and Chandigarh have undertaken reforms partially by amending APMC Act and the States of Mizoram, Meghalaya, Haryana, J&K, Uttrakhand, West Bengal, Puducherry, NCT of Delhi and Uttar Pradesh have just initiated administrative action for the reforms.

Status of Agri-Market Reforms

Sl. No.	Stage of Reforms	Name of States/ Union Territories
1.	States/ UTs where reforms to APMC Act has been done for Direct Marketing; Contract Farming and Markets in Private/ Coop Sectors	Andhra Pradesh, Arunachal Pradesh, Assam, Chhattisgarh, Goa, Gujarat, Himachal Pradesh, Jharkhand, Karnataka, Maharashtra, Mizoram, Nagaland, Odisha, Rajasthan, Sikkim, Uttarakhand and Tripura.
2.	States / UTs where reforms to APMC Act has been done partially	a) Direct Marketing: NCT of Delhi, Madhya Pradesh b) Contract Farming: Madhya Pradesh, Haryana, Punjab and Chandigarh c) Private market: Punjab and Chandigarh
3.	States/ UTs where there is no APMC Act and hence not requiring reforms	Bihar*, Kerala, Manipur, Andaman & Nicobar Islands, Dadra & Nagar Haveli, Daman & Diu, and Lakshadweep.
4.	States / UTs where APMC Act already provides for the reforms	Tamil Nadu
5.	States / UTs where further action is required for the reforms	Meghalaya, Haryana, J&K, West Bengal, Puducherry, NCT of Delhi and Uttar Pradesh.

*Bihar has repealed APMC Act

Impact of Market Sector Reforms as per Model Act- From above chart it may be seen that the level of adoption of Model Act is varying, but it is noteworthy that mere adoption of Model Act has neither resulted into enhanced investment of private sector in value chain management for fresh horticulture produce nor in a better price discovery under process of direct procurement of agriculture produce under contact farming system. The Working Group has critically examined following case studies for this purpose:

a. Status of F&VMarketing in Bihar after abolition of APMC Act-after the APMC Act has been repealed in the State of Bihar, w.e.f. September, 2006; market infrastructures created earlier by the Bihar State Agricultural Marketing Board in the State are presently in use by the trader who were earlier operating from their shops allotted to them on rent; Sub Divisional Magistrate of the area functions as the Nodal Officer and he is in-charge of the unregulated markets and no market fee are charged from the farmers. However, other charges towards loading/ unloading/ Hamal charges are in vogue. But this reform has not resulted into enhanced private investment for development of market infrastructures for agriculture produce in the State; as such the State of Bihar has gone back to primitive stage of market infrastructure in respect of agriculture produce.
b. Organized retail chain outlets for fresh fruits & vegetables by corporate Sector- Earlier the public sector organisations like NDDB and the Horticultural Producers' Co-operative Marketing and Processing Society Ltd. (HOPCOMS) had taken initiatives to organize marketing of horticulture produce and set up retail outlets in cities in early eighties, however, these experiments had recorded only limited success in terms of establishing backward linkage with producer farmers and promoting formation of common facility centres in producers' clusters. After the Market Sector Reforms in agriculture/horticulture, a number of corporate houses have set up organized retail chain outlets in big cities but they too, generally function as wholesale buyers of horticulture produce and lack backward organic linkage; they do not provide any price discovery mechanism to producers which are better than APMC system. Even in terms of commodity price and quality these outlets do not make any material difference either for producers or for consumers.
c. Terminal Markets-Terminal Markets Complex (TMC) introduced as a new item under NHM is conceived to be set up in a Public Private Partnership (PPP) mode. It is designed to be set up in the form of hub and spoke model. It has provision of equity participation by Producers Association up to 26% of the total equity in the TMC.

Approval has already been accorded for establishment of TMC at Patna(Bihar) and Perundurai (Tamil Nadu) and in-principle approval has been accorded for bidding for TMC at Babangaon (Maharashtra), Nagpur (Maharashtra), Madurai (Tamil Nadu), Kancheepuram(Tamil Nadu) and Sambalpur (Odisha). So far, the mega-scheme of Wholesale Market and Terminal Markets under National Horticulture Mission has not picked up during 11th plan period in spite of viability gap funding to the extent of 50% of project cost extending to Rs. fifty crores.

Recommendations of a Committee of State Ministers in-charge Agricultural Marketing

In order to expedite the pace of market reforms the Ministry of Agriculture has set up a Committee of State Ministers In-charge Agricultural Marketing on 2nd March, 2010, with members from the state of Maharashtra, Gujarat, Haryana, Uttarakhand, Bihar, Assam, Orissa, Andhra Pradesh, Karnataka and Madhya Pradesh. The report called for "speedy reforms" of Agricultural Produce Market Committees Act across different states along with "time-bound development" of agri-marketing infrastructure. Calling for a 10-year perspective plan to improve agri-infrastructure of backward and forward linkages for agriculture production and marketing, the report suggested providing "priority sector lending to agri-marketing activities".

As the initiative taken for Market Sector Reforms have not yielded desired results, perhaps that is why Approach Paper to the Twelfth Five Year Plan (2012-17) has echoed the need for excluding horticulture produce and perishables entirely from the ambit of APMC, as a viable solution.

Approach Paper to the Twelfth Five Year Plan (2012-17): The Approach Paper to the 12th plan states that "Consideration should be given to extending infrastructure status to a wider range of agricultural market facilities in the same manner as for warehouses. States must modify the Essential Commodities Act (ECA) and the APMC Act (perhaps exclude horticulture and perishables entirely from the ambit of APMC), rebuild the extension system, increase the involvement of private sector in marketing, and also facilitate leasing in/out of land by farmers."

Critical Analysis of Suggested remedy in Approach Paper to the Twelfth Five Year Plan: The Working Group has critically examined the implications of the suggested remedy in Approach Paper to the XIIth Plan as mentioned above. Taking horticulture produce out of ambit of APMC would mean that it may not be mandatory to trade them in any market premise, be it public or private, therefore; it could not be comprehended as to how private sector investment would materialise in marketing infrastructures if horticulture produce is taken out of ambit of APMC. In this context the case of flower markets may be taken into account which is not the scheduled commodity under APMC Act and to that extent excluded from ambit of APMC. In spite of this, the flower-markets are in no different shape than markets of any other horticulture produce. By taking horticulture produce out of ambit of APMC may have three implications; firstly, this would result into Bihar like situation which prevails after abolition of APMC Act or situation of flower market of Mumbai which takes place below Dadar flyover or Delhi flower market which takes place near Hanuman Temple in Connaught Place. Secondly, it would mean, direct procurement from farmers' field by operators of organised retail chain outlets/processors and exports which alone will not be able to introduce a price discovery and transparency regime in trade of horticulture produce and may not be able to deal with huge volume of horticulture produce in our country. Moreover, from the other important recommendation made in the Approach

Paper to XIIth plan regarding leasing in/out of land by farmers may ultimately mean promoting corporate farming model which would have far reaching socio-economic implications that cannot be ignored after having experience of land acquisition for corporate through the instrument of Land Acquisition Act. Permitting Corporate Farming and relaxation of ceiling laws may not even remotely really address the issue of Market sector reforms but may have serious taxation related implications are high.

- State of Unregulated Rural Primary Markets/Haats/Shandies/Painths: At present, horticulture produce along with other farm output is traded through a network of about 27738 Wholesale and Rural Primary Markets out of which only 7157 are Regulated Markets. Other Rural Primary Markets include Haats, Shandies, Painths and fairs which are estimated to be more than 20,000 across country which are owned and managed by the private individuals, Panchayats and municipalities who are interested in collection of ground rent/fee/cess. As such, generally revenue realization from market is hardly applied for the development of infrastructure in Rural Primary Markets. About 90 percent of the total marketable surplus in the remote areas is sold through these markets. It is assessed that the efficiency of rural markets is poor due to high degree of congestion at market yards, less number of traders and non-availability of supporting facilities and services and private sector investment is not coming for Primary Rural Markets.

- State of Affairs at unregulated standalone wholesale markets: Like Primary Rural Markets majority of standalone wholesale markets handle mix of commodities including, horticulture produce. Infrastructure facilities available in these markets at present are far from satisfactory and do not cater to special needs of perishables. In addition, there are some special types of unregulated markets as given below-
- Unregulated Wholesale Markets in Kerala Managed by Public Authorities: As such there is no market regulation in Kerala, there are 6 Agricultural Wholesale Markets in Kerala under the control of State Agricultural Department, out of which 3 are Urban Whole Sale Markets and 3 are Rural Wholesale Markets. The Agricultural Urban Wholesale Market in Maradu, is one of the main wholesale market in Ernakulum district. The market authorities charge rent for the stalls from the traders. The traders buy produce from farmers and also bring from other states. Farmers bring their produce to the market once in a week and the traders buy directly from them. At present this Market is not charging anything from the farmers or traders for this service, to that extent, this model is subsidised my Government for its operating expenses.
- Unregulated Wholesale Markets in Kolkata Managed by Private Sector: Kolkata city has unregulated private wholesale markets which are called as Koley Market for vegetables, Machua Market for Fruits or Mallighat Flower market which are managed by private persons who run the same in the most primitive manner and charge market fee based on weight of produce. From the following photographs, glimpse of the state of affairs at these privately managed wholesale market of Kolkata can be seen-
- Case of APPTA Market in Nagercoil (TN): Another type of unregulated market had been set up and run by Agricultural Products Producers and Traders Association (APPTA) in Tamilnadu which is also known as APPTA Market. APPTA Market is a Modern Fruit and Vegetables Market constructed at Nagercoil near Kanyakumari in Tamil Nadu. It is the important major assembling centre for fruit and vegetable in a radius of 50 km. The infrastructure facilities provided in the markets are wholesale shops (131), retail shops (504), covered auction hall, open auction platforms, storage godowns, pre-cooling/ripening chambers and drying yard. Input shops for fertilizer, seed, pesticides and grocery shops are also constructed in the market complex. Provision is made for other public utilities like drinking water, drainage, toilets, post office, bank, internet kiosk,police out post, bus stop, famers guest house,

tea shops, hotel and restaurant. Appropriate handling capacity of the market is 3,000 MT of Fruit and Vegetables per day. The arrivals are reported from within the district and neighbouring Tirunelveli and Tuticorin districts in Tamil Nadu. Dispatches are mainly to Kerala markets and some quantity to the northern districts of Tamil Nadu, Chennai and Bangaluru. The revenue for the market is from entrance fee, rent and maintenance charges. The unexpected risk factors are huge investment but lesser return, higher land cost, high interest burden, less amount of subsidy and very important lack of government support. Due to these reasons the market is financially weak and they are unable to repay the bank loan.

- Modern Wholesale Markets with Sorting Grading & Electronic Auction Facilities: Wholesale market has been set up by NDDB in Bangaluru city with technical assistance from the FAO. It is based on the European model of wholesale trade. The facility comprises state of the art auction halls with computerized bidding, specially designed cold stores for different produces, shops and other related commercial and administration facilities. The superstructure is entirely in steel with a cladding of PUF panels. NDDB has taken initiative to organise producer farmers as SAFAL-Groups which could have addressed to the issue of consistency in quality and traceability of produce for the purpose of processing and exports. Management of these markets own function of market management, facilitate business model which ensures price discovery, protection to sellers regarding realization of sales proceeds and facilitates to buyers the quality of produce. However, capacity utilisation of this market is sub optimal due to non-participation of traders, processors, exporters etc and its business is generally limited to local retailers who source their commodities from retailer counter of this market .
- Modern Flower Auction Houses set up under Assistance from APEDA: four modern flower auction centres have been set up by respective State governments under financial assistance from APEDA as per design and business model recommended by consultants of international repute. These auction centres not only provide sorting & grading facilities, cold storages, electronic auction system but also banking facilities, offices to producers and traders and value addition facilities. Management of these markets own function of market management, facilitate business model which ensures price discovery, protection to sellers regarding realization of sales proceeds and provides quality assurance of produce to the buyers.

Inference regarding Present Strategy of Market Sector Reforms

a. Rural Primary Markets may not attract sizable private sector investment in its infrastructure and public sector investment will be required. There are schemes of Gramin Bhandaran Yojana, Development/Strengthening of Agriculture Marketing Infrastructure, Grading and Standardisation implemented by Directorate of Agriculture Marketing and Inspection, DAC, scheme-component under NHB of giving Assistance for setting up sales outlet for horticulture produce and scheme of assistance for setting up Rural Primary Market under NHM and MM-II of HMNEH which need to be continued during XIIth plan period.

b. Even in cases of wholesale market segment, investment in market infrastructure in respect of a large range of horticulture produce may not generate handsome returns and therefore, expecting private sector investment in development of majority of wholesale markets may either not materialise or may not be of the standards which is targeted by such initiatives for market reforms. The reasons behind poor take off of schemes of Wholesale Market and Terminal Markets offered by NHM inspite of sizable viability-gap funding appears to be linked with absence of Standard Operating Procedure (SOP) for traditional Horticulture Wholesale Markets which means absence of level-playing ground for the investors in modern marketing infrastructures with SOP. This is reinforced by the experience of yet another model of Public Sector Investment in infrastructure combined with professional management by stake-holders which are exemplified by NDDB's Wholesale F&V Market at Bangaluru and Modern Flower Auction Houses set up with assistance of APEDA.

c. To conclude, Market Sector Reforms without having Common Standard Operating Procedure (SOP) Strategy for Addressing other Important Issues of Marketing and Prices of Horticulture Produce?

Bridging wide gap between wholesale and retail prices of horticulture produce in urban areas- Producers' Market in Towns/Cities-Existence of a multi-layered middle-men in value chain management of perishables and higher post harvest losses have been cited as the main cause of wide gap between wholesale and retail prices of fresh horticulture produce and case of introduction of organised retail chain outlets has been advocated to improve the efficiency of value chain. However, in practice, the existence of NDDB's SAFAL outlets and retail chain outlets operated by players of corporate sector have not been able to present any better model in cities like Delhi, Bangaluru, Mumbai etc. On the other hand, direct farmers' market have been in existence in many towns and cities; for example- Pool Gate Market of

Pune Cantonment, Vasantkunj market of New Delhi and Anta Ghaat Market in Patna town are some of the examples of age- old, existing farmer's markets. These were phased out by regulated markets under concerns of town planning, traffic regulation and economic viability of market management. In recent past, the direct farmers' market for perishable got reintroduced through a few innovations carried out for facilitating direct marketing of agriculture produce including horticulture produce by the producer farmers. Some of these innovations are named as Uzhavar Sandhai introduced by Tamil Nadu in 1999-2000 for direct selling of fruits and vegetables by farmers to consumers at a fair price, Rythu Bazaars in Andhra Pradesh to provide direct link between farmers and consumers in the marketing of fruits, vegetables and essential food items, Apni Mandi in Punjab, to establish direct contact between the farmers and ultimate consumers for sale of the produce and Krushak Baazars have been established by State Government of Odisha and are managed by APMCs. Farmers generally trade paddy, maize and cotton and fruits and vegetables in Krushak Baazars. This leads us to infer that modern town planning and traffic regulation should take into account the relevance of direct farmers' markets in respect of perishables which may go a long way in reducing the gap between wholesale price and retail price. However, this may address to the issue of marketing needs of only a section of consumers which is more sensitive to price of the produce and is willing to go to such markets operating at certain distance from their residential areas.

Market Information System and Market Intelligence- At present, Market Information in respect of agriculture including horticulture produce is collected by Directorate of Agriculture Marketing, DAC, the Government of India through AGMARKNET Project. Under this project, market price and arrival is collected by mandi staff and uplinked on AGMARKNET portal. 'AGMARKNET' is a unique live portal on agricultural commodities, technically supported by a high capacity Central server and the programming capabilities of the NIC and the data is fed into the system and later disseminated to farmers in a decentralized mode through the voluntary cooperation of mandi staff. In addition, NHB too collects information regarding market price and arrivals from selected markets and uplinks the same on its own website. The NHB also releases monthly and annual market bulletins. In addition, NHB publishes annual horticulture database too by way of marketing related information. The efforts of DMI and NHB may be coordinated by making later responsible for validation of data. Even, the newly introduced system of Direct Procurement of horticulture produce by Managers of Organised Retail Chain Outlets, Processors and Exporters has not resulted into better price discovery and a better market information system. In the present scenario it is not possible to get real-time, truthful data regarding commodity arrival and price for agriculture/horticulture produce and generation of Market Intelligence remains a distant reality. Introduction of SOP can facilitate generation of real-time, truthful market information and market intelligence for horticulture produce.

Policy Interventions- Working Group is of the view that excluding horticulture produce and perishables entirely from ambit of APMC may benefit only those interested in business of operating organized retail chain outlets or sourcing the raw material for processing but it will prove to be counterproductive in terms of development of market infrastructures. In such circumstances, expecting any kind of private investment in market infrastructure free from conflict of interest will be a dream as no investor will make investment in market infrastructure if he is not sure of business inside market premise. Therefore, the Working Group has made following recommendations which are more in consonance with the ides of properly designed Market Sector Reforms.

Redefining Role for Management of Regulated Markets and Introduction of Standard Operating Procedure (SOP): It is proposed that the management of regulated markets should also be compelled to follow standard Operating Procedure (SOP). This will not only improve the functioning of existing regulated markets which will be compelled by business requirements to make investment in packing& grading, modern ware houses and electronic auction system but introduction of Standard Operating Procedure (SOP) will pave the way for investment in Private Markets.

Rationale Policy of Registration of Wholesale Buyers-As per present practice, in a regulated market, the wholesale buyer is licensed by Market Committee. Once a system of registration of wholesale buyer under SOP is introduced, such buyers will be able to participate in auctions to the extent of bank credit limit furnished by them to the market. Therefore, introduction of new wholesale buyers should be encouraged and a rational procedure in this regard needs to be introduced.

Permitting setting up Private Markets in Area of Operation of APMCs- It is proposed to permit setting up of private markets in area of jurisdiction of AMPCs so that there is a healthy competition for providing better services. Market Regulator should ensure that those involved in management of Private Markets do not have conflict of interest. Private Markets and APMC should be subject to same Standard Operating Procedure (SOP).

Trends in Area, Production and Productivity

Indian cardamom is one of the exotic and highly priced spices. Special cultivation and processing methods combine to make Indian cardamom truly unique in aroma, flavour, size and colour-tempting parrot-green. There are two types of cardamom grown in India, viz. small and large cardamom. There are three varieties of small cardamom grown in India, viz. Malabar, Mysore and Vazhukka.

Malabar: It is medium in size and attains 2-3 meters height at maturity. The panicles are prostrate and the capsules are globose to oblong shaped. This variety is better suited to 600-900 meter elevation and is mostly cultivated in Karnataka, Lower Pulnies of Tamilnadu and Kannielam tract of Kerala.

Mysore: It is robust and attains 3-4 meters height at maturity. The panicles are erect and the capsules are ovoid bold and dark green in colour. It is better adapted to altitudes ranging from 900 to 1,200 meters. This is mostly cultivated in Kerala and in certain pockets of Tamil Nadu and Karnataka.

Vazhukka: It is considered to be the natural hybrid of Malabar and Mysore types. The plants are robust and the capsules are bold globose or ovoid in shape. It is extensively cultivated in Kerala and Tamil Nadu at elevations ranging from 900 to 1,200 meters above msl.

However there are new selections made by growers. The most widely diffused one in Kerala and Tamil Nadu with much higher yield is called Njallani Gold (after the name of the grower who developed it). There are also other selections made by growers. Conventionally cardamom has been cultivated in evergreen forest after cleaning the undergrowth and selectively thinning the tree canopy for shade regulation. The absence of smaller trees in the cardamom estates suggest that there are difficulties in forest regeneration when the over storey trees die (Kumar et al. 1995). Hence in a context of high environmental consciousness, conversion of natural forest into cardamom cultivation is considered as inimical to environmental sustainability. Therefore, the observed decline in area under cultivation is a natural outcome in a context wherein there are obvious limits to bringing new areas under cultivation by clearing forest cover. In this context it has been proposed that mixed cropping practices are more sustainable using agro-forestry methods for cardamom cultivation (Singh et al. 1989; Sharma et al. 1994; Rao et al. 2004). A dense multiple canopy and inter-cropping of cardamom with a variety of other crops and trees have been proven ecologically and economically sustainable and it could even increase the fertility of already depleted soils (Moench, 1991). Here it is worth noting that a reorientation towards mixed cropping in districts like Idukki with planting new trees (preferably the ones that bring an additional income like Jack? Mango?

Brazil fruit Mangostein?) for ***Kerala and Towards a Strategy*** shade provision. This will also enable the cardamom growers to claim carbon credit under the Kyto-protocal. Thus there lies an opportunity to reverse the observed trend in area under cultivation. If this proposition is accepted there is a need to reorient the ongoing strategy of replanting/replanting programmes of the Species Board as well as the co-ordination with the Ministry of Environment and Forest.

Share of Yielding Area to Total Area of Cardamom across States and India

Year	% of Yielding to total Area (at all India)	% of Yielding to total Area (Kerala)	% of Yielding to total Area (Karnataka)	% of Yielding to total Area in Tamil Nadu
1990-91	75.09	78.13	71.58	71.52
1991-92	76.77	79.22	74.48	71.24
1992-93	74.20	73.49	76.80	65.29
1993-94	73.34	74.73	72.25	69.48
1994-95	74.03	73.51	75.49	69.64
1995-96	73.84	73.59	74.72	70.66
1996-97	71.84	74.17	68.97	68.40
1997-98	73.04	75.04	70.33	70.92
1998-99	73.36	74.17	71.38	76.76
1999-00	74.08	74.23	73.45	76.07
2000-01	74.22	74.27	74.50	72.37
2001-02	74.37	74.60	74.42	72.16
2002-03	74.67	74.51	75.21	73.21
2003-04	75.40	74.98	76.42	73.42
2004-05	75.33	75.43	76.56	68.19
2005-06	74.78	74.29	76.94	67.40
2006-07	73.94	73.06	76.56	67.55
2007-08	71.36	71.18	71.45	72.38
2008-09	70.89	69.97	72.47	70.67

Trends in Area under Cardamom across States

Year	Kerala	Annual Growth Rate of Area in Kerala	Karnataka	Annual Growth Rate of Area in Karnataka	Tamil Nadu	Annual Growth Rate of Area in Tamil Nadu
1990-91	43826		31605		6123	
1991-92	43670	-0.4	32052	1.4	6123	0.0
1992-93	43388	-0.6	32881	2.6	6123	0.0
1993-94	43459	0.2	33414	1.6	6087	-0.6
1994-95	44237	1.8	33518	0.3	5896	-3.1
1995-96	44248	0.0	33743	0.7	5811	-1.4
1996-97	41268	-6.7	25996	-23.0	6329	8.9
1997-98	40867	-1.0	25686	-1.2	5891	-6.9
1998-99	41449	1.4	25638	-0.2	5048	-14.3
1999-2000	41491	0.1	25882	1.0	5056	0.2
2000-01	41288	-0.5	25947	0.3	5085	0.6
2001-02	41336	0.1	26258	1.2	5069	-0.3
2002-03	41412	0.2	26644	1.5	5069	0.0
2003-04	41332	-0.2	26838	0.7	5067	0.0
2004-05	41378	0.1	27094	1.0	5253	3.7
2005-06	41367	0.0	27173	0.3	5255	0.0
2006-07	41362	0.0	26611	-2.1	5255	0.0
2007-08	39763	-3.9	24976	-6.1	4561	-13.2
2008-09	41588	4.6	25021	0.2	4561	0.0
Trend Growth Rate	-0.18		-0.66		-0.7	

Trends in Yield of Cardamom across States

Year	Kerala	Annual Growth Rate	Karnataka	Annual Growth Rate	Tamil Nadu	Annual Growth Rate
1990-91	101		35		113	
1991-92	102	1.0	44	25.7	115	1.8
1992-93	81	-20.6	48	9.1	120	4.3
1993-94	136	67.9	63	31.3	156	30.0
1994-95	145	6.6	62	-1.6	170	9.0
1995-96	165	13.8	69	11.3	189	11.2
1996-97	149	-9.7	76	10.1	165	-12.7
1997-98	172	15.4	103	35.5	180	9.1
1998-99	162	-5.8	83	-19.4	173	-3.9
1999-2000	214	32.1	102	22.9	208	20.2
2000-01	247	15.4	109	6.9	217	4.3
2001-02	272	10.1	108	-0.9	238	9.7
2002-03	281	3.3	115	6.5	251	5.5
2003-04	286	1.8	85	-26.1	259	3.2
2004-05	276	-3.5	91	7.1	257	-0.8
2005-06	318	15.2	85	-6.6	282	9.7
2006-07	283	-11.0	85	0.0	272	-3.5
2007-08	248	-12.4	89	4.7	253	-7.0
2008-09	294	18.5	93	4.5	233	-7.9
Trend growth rate		3.00		1.27		1.41

Processing and Marketing of Large Cardamom in Sikkim and North East Region

Processing

After harvesting, large cardamom is collected and transported to curing units and prepared for storage. Freshly harvested large cardamom contains around 70–80% moisture content, which must be reduced to 10% (maximum) in order to store it properly without rotting (Zala 2002). The traditional curing unit (kiln), or 'bhatti' as it is called locally, is a simple curing device made from materials found locally, such as stone and mud, with a wire mesh layer sitting above an open chamber. A fire is lit below the shelf and the heat absorbed in a stonewall or heat sink, which radiates the heat and dries the capsules (Rao et al. 2001).

However, the various inadequacies of this system make traditional bhattis less than desirable for producing a marketable end product. Drawbacks include uneven heating and drying; loss of volatile oils in the cured capsules (Zala 2002); loss of the characteristic rich, dark red colour; charring; and an infusion of smokiness from the wood fire below (Rao et al. 2001). Recognizing these drawbacks, the Spices Board of India has developed a modified bhatti with a 200 kg or 400 kg capacity. This modified bhatti uses heat that is more evenly distributed to cure the cardamom and generate a higher quality product and is offered at a subsidized rate by the Spices Board under the scheme 'Export Oriented Production and Post Harvest Improvement of Spices' (Spices Board 2011).

Marketing

Once the large cardamom capsules are cleaned, cured, and graded there are several avenues that farmers can pursue for marketing. Many farmers sell their product directly to local merchants or those in nearby towns. In exchange, the merchant sets up an account for the farmer, which the farmer can use to buy provisions for the following year. The main drawback of this practice is that farmers are at the mercy of the merchants and are offered less than the actual value of the large cardamom. It is also common to sell produce through a middleman. The middleman visits each farm in a locale, buying and collecting the cardamom from each farmer along the way. The middleman then sells the product either to local merchants or wholesalers in Siliguri. This method is prevalent in areas with little or no access to roads and with mostly small-scale farmers who have neither the time nor the money to bring their produce to market. Here, the farmer is at a disadvantage as s/he must provide the cardamom at the rate that the middleman proposes to pay, which is generally less than s/he would get directly from the merchant. Once in the hands of the local or external wholesaler, the cardamom is sold at a higher rate on local, national, and international markets. While India continues to be the largest consumer of large cardamom in the world, a huge percentage is also sold internationally. The second largest consumer of large cardamom is Pakistan, followed by United Arab Emirates, United Kingdom, United States of America, Canada, and South Africa (Maheskumar 2010).

Since 2010, a third marketing channel has been developed by the North Eastern Regional Agricultural Marketing Corporation, whereby auctions are held monthly in Rangpo, a nearby market town. This large cardamom auction has increased the return to farmers by Indian rupees (INR) 25 (USD 0.46) per kg (Personal communication with North Eastern Regional Agricultural Marketing Corporation on 11 February 2013) by creating a competitive bidding process between merchants. The record price one farmer received was INR 1,240 (USD 23) per kg. The auction also incentivizes farmers to improve their cultivation and post-harvest processing methods. The danger of receiving too low an offer is eliminated by giving the farmer the opportunity to withdraw his or her submission, even after the bidding has taken place. Plans for the development of a new 'E-auction' are underway. An E-auction would improve the situation in many ways, including by facilitating international buyers to purchase cardamom and creating even more competition among potential buyers, resulting in a higher return for farmers. However, if large cardamom is made available to the global market, Indian merchants could easily be outbid and the price of cardamom would increase for Indian consumers. In 2010/11, the North Eastern Regional Agricultural Marketing Corporation sold 4 metric tonnes of cardamom through an open auction. The following year, the sale grew to 16 metric tonnes. In the year 2013, it is expected to exceed 40 metric tonnes. The average auction price for cardamom in 2010 and 2011 was INR 1,100 (USD 20.4) per kg, which is expected to grow to INR 1,300 (USD 24) per kg in 2013. The North Eastern Regional Agricultural Marketing Corporation operates the only large cardamom auction centre in India at Rangpo in Sikkim. The marketing system for large cardamom in Sikkim is improving, however the remaining produce of large cardamom (>3,000 metric tonnes) is sold without organized marketing channels, thus resulting in a minimum premium priceto farmers

Pollination of Large Cardamom

Main Pollinators

Large cardamom is essentially a cross-pollinated crop and requires the cross-pollination of its flowers for crop production. Different species of bees are its main pollinators. However, despite its economic importance, there are very few studies on the pollination eco biology of this crop. Preliminary research by Verma (1987) reported the honeybee, *Apis dorsata*, to be one of the pollinators of large cardamom by directly relating the increase in cardamom production to an increase in the number of *Apis dorsata* colonies. However, in recent years, interest in cardamom pollination is growing among the scientific community, resulting in an increase in information about its pollination requirements, pollinators, and the impact of pollination on cardamom production. A few studies have reported that common native species of bees including *Apis cerana*, *A. dorsata*, *A. florea*, and *A. laboriosa*, as well as the less noticed stingless bee, *Trigona* sp. (putka), visit large cardamom flowers (Singh et al. 2012). Further studies by Sinu et al. (2011) reported that *Bombus haemorrhoidalis*, *Apis cerana*, *Megachile lanata*, *Episyrphus balteatus* (hoverfly), *Macroglossum stellatarum* (hawk moth), and *Aethopyga siparaja* (crimson sunbird) also visit the flowers of large cardamom and that *B. haemorrhoidalis*, *M. lanata*, and *A. siparaja* were the most effective pollinators of large cardamom. Deka et al (2011) reported that two bumblebees, *Bombus haemorrhoidalis* Smith and *B. breviceps* Smith were the predominant pollinators of large cardamom.

Role of Apis cerana

A recent study on large cardamom pollination by Sinu and Shivanna (2007) showed that 70% of its flowers visited by the indigenous honeybee *Apis cerana* remain unpollinated, affecting the cardamom yield. These authors further reported that *Apis cerana* collects a large portion of the pollen from the flowers of large cardamom leaving less chance for pollination by other pollinators. However, *Apis cerana* beekeeping is an age-old traditional practice and an integral part of traditional farming system in Sikkim. The local people keep native species of honeybee in log or box hives made of wood hung against the wall. A few households also keep bees in wall hives. The findings of the study revealed that nearly two-thirds (64%) of the respondent households engage in beekeeping. In addition to beekeeping, honey hunting is also practised in Sikkim, especially in forest areas and cliffs. Thus, it is common to find the indigenous bees co-existing with cardamom farms in many villages. Large cardamom farmers in the study sites were asked whether or not they have experienced a decline in cardamom production because of *Apis cerana* beekeeping in their villages. Almost all (99%) of the respondents said that bumblebees and honeybees pollinate large cardamom. Around 59% of cardamom growers also observed that honeybees are important agents of pollination. Over 70% of respondents reported that honeybees do not affect cardamom production negatively, a small number (around 7%) said that honeybees do affect cardamom bproduction negatively as they rob them of pollen, and 23% had no idea whether or not honeybees have a positive or negative effect on cardamom production. Interestingly, none of the key informants mentioned that honeybees negatively affect cardamom production. Ninety per cent of key informants and 71% of cardamom growers expressed the view that honeybees are not responsible for the decline in cardamom production. Interpreting this information from farmers, it is possible that, even though under control conditions (i.e., when there is only *Apis cerana* visiting cardamom flowers and other pollinators are absent) *Apis cerana* may not be a very effective pollinator of large cardamom flower due to its small body size, it does not appear to have a significant negative impact in the open field as reported in earlier studies (Sinu and Shivanna 2007). There are a large number of other pollinators that help in pollinating cardamom under open field conditions; however, further research isneeded to verify this. One thing that is very clear from the present study is that it is very important to conserve and protect the major cardamom pollinators, namely, bumblebees, as in their absence farmers may not be able to harvest cardamom of an optimal quantity and quality. The majority of farmers who responded to the survey (74%) and almost all of the key informants cited bumblebees as the most important pollinators of cardamom, a finding that confirmed the scientific studies (Sinu and Shivanna 2007; Deka et al. 2011). Additionally, as many as 88% of cardamom growers and 65% of key informants said that they have seen honeybees in the wild and in their surrounding areas.

Farmers' Perceptions of Large Cardamom Pollination Under Climate Change

All of the key informants (100%) and 83% of the cardamom growers reported a serious decline in pollinating insects and a consequent decline in cardamom production over the years. The majority of key informants (90%) and cardamom growers (70%) believe that this decline is due to the impacts of climate change. Although it appears that much of this decline is due to a lack of pollination resulting from climate change and a decline in pollinator populations, how much of it is due to other factors could not be established. However, it is clear from the study that climate change is negatively affecting populations of insect pollinators. Around 15% of key informants and 80% of cardamom growers categorically observed and confirmed that the rising temperatures, long dry spells, and fluctuating rainfall patterns have resulted in the transformation or destruction of the micro-habitats of insects and altered their breeding patterns. Farmers believe that the decrease in populations of insect pollinators is also due to the change in cropping patterns, land use change, and unregulated developmental activities such as road construction, building construction, hydropower projects, pollution, and the accumulation of dust particles in the air Climate change-induced changes in local weather conditions have been shown to increase the incidence of pests and diseases and affect pollinator populations and pollination in various crops (Partap and Partap 1997, 2002, 2009). Climate change is reported to affect the flowering patterns of various crops and plants and the availability of pollinators (Rana 2010; Sharma and Partap 2011), and to shift cropping patterns in the Himalayan region (Bhatt 2010). Respondents have seen the nests of wild bees hanging on the cliffs and once a year some villagers collect honey from these bees. This honey is considered to have high medicinal value and is distributed among village households. A few farmers also keep stingless bees called 'putka' (*Trigona* sp.) in their houses. The honey from putka is believed to have great medicinal value and is sold at USD 112 per kg.

Current Scenario on Production and Marketing

2014 monsoon is almost nearing its withdrawal period and the entire country is recently facing a deficiency of 18% rainfall. The late revival of monsoon has affected the acreage of many crops across the country. However, some crops with a late sowing window were blessed with monsoon showers wherein a likely increase in acreage is witnessed as farmers have shifted the area to late sown crops. Crops such as Cotton and Rice have crossed previous year's sowing area as on date. Meanwhile, crops like cardamom were affected adversely during the initial part of monsoon owing to scanty rains. Recently descriptive research conducted a primary survey of Cardamom in the state of Kerala and Tamil Nadu during the second and third week of August. The observations seem to be very disappointing for the commodity. Rough estimation suggests, this year cardamom is grown in less than 26,000 hectares across the country while the preliminary estimation shows production may drop by 30-40% from its previous year's total supply of 18500-19000 MT (14,000-14500 MT production plus 3,500-4500 MT carry forward stocks). This effectively means actual final production in the 2014 could stand around 12,000-12500 MT. If we consider such figures then the commodity is likely to face a deficit situation that may drive the cardamom prices northwards.

Therefore, we need to reiterate the facts and figures and delve a fresh outlook on the commodity, likely export trend, consumption pattern, the estimated carry forward stocks and finally the price potential on the higher side. The survey suggests that first picking of the cardamom is completed during the initial week of August and the yield so far has been less than 25% of India's production. Also the harvested crop is of small grade quality and the approximate production figure stays subdued around 3,000 to 3,500 MT. We understand that inferior crop quality may keep the export demand for the commodity soft which indirectly might keep cardamom prices in a moderate trend. Historically, it has been noticed that no export consignments are made during the month of August so possibly the price trend may remain subdued in the short-term. As of last week of August, cardamom prices are trading near Rs. 800 per Kg and the trend may likely remain steady to slightly bearish. However, as we proceed towards September month, second picking is likely to start. Usual interval between two pickings stands around 40 days and we expect the quality in the forthcoming second picking season to improve wherein a higher grade may be obtained. We keep our finger crossed as scanty and scattered rains might have spoiled the growth status of the commodity and possibly because of which the supply which are anticipated to be good in September mid may be disappointing. If such scenario props up then possibly a further shortfall in the crop supply may be seen that may push the prices higher. Nevertheless, if normal scenario persists then possibly good supply might pull the prices a tad lower during the month of September. As per the seasonality index, it is noticed that Cardamom prices tend to remain bearish until September and starts to gain from October. As discussed above, for this year we have to keenly watch the likely arrival pattern and its impact on the prices. Also, we expect that supplies which come during September and October would be of better quality i.e. 6 to 7 mm grades of cardamom. It is likely possible that good quality cardamom would enhance the prospects for exports that too might help the commodity to move northwards. Thereafter, at the end of November or early December the last picking takes place which are generally of lower quality produce. However during that phase, higher winter demand may support the prices to continue trade higher.On a cumulative note, we believe Cardamom may remain in deficit in 2014 and that might keep the commodity prices elevated over the short to medium term. Note that traders and investors need to closely watch the second picking status and its effect on the prices. Looking at the seasonal pattern, we believe if prices make a slight dip during September then physical buying could drive prices higher. From the global front the largest cardamom producer i.e. Guatemala is likely to produce around 25,000+ MT this year which is precisely of less quality crop. On that regard too, good competitive advantage may be noticed on Indian cardamom due to better grade. Overall, we expect cardamom prices may remain bullish wherein we recommend buying the same on small corrections. On a pricing perspective, we advice taking long positions from around Rs. 850/Kg (spot) levels where an appreciation of 40% to 45% is likely by the month of February / March 2016. On the lower side the risk remains within 10%.

System of Innovation in Cardamom Marketing

Cardamom is a plantation crop grown historically in the evergreen forests mainly for export3 and its price depends on the quality of the dried capsule. The quality of the product in turn depends on the colour, aroma, and boldness of the capsule. All these depends on the plant variety, timing of harvest (which is a highly skilled and women labour intensive) curing method adopted (sun drying or in curing houses) and timely implementation of cultural operations especially spraying of pesticides to protect the capsules from a number of insects and pests. To the extent that there is significant variation across and within holdings in terms of harvesting, curing and other cultural operations along with plant variety there is significant quality variation across and within output offered for sale by a given grower and across different growers. Given such quality variation an auction system has been followed historically for the marketing of cardamom4. Until the introduction of e-auction in 2009 in Bodinaikanur (Tamil Nadu) and 2010 in Puttadi (Kerala) auction has been in the traditional outcry method wherein the bidders competed face to face for the lots offered for sale.

The nature of regulation is found to be varying from crop to another. In case of cardamom, the regulation takes the form of restricting the entry of different functional categories in the like of the sellers, auctioneers, dealers, and exporters into the auction system as per the Cardamom (Licensing and Marketing) Rules of 1977. The declared objective of such a regulation has been to ensure a fair price and timely payment of sale proceeds (Nair et al 1989, Narayana 1994, Joseph 1985).

The marketing system for cardamom market has been characterized by large number of sellers (both large and small) a small number of exporters – many of them are dealers and few of them, in addition to being dealers, are growers as well – a large number of dealers and a few auctioneers. Auctioneers are one of the major actors in the marketing of cardamom who bring together the buyers and sellers of the product. During 1980s when the auctions were conducted in the conventional outcry manner and held in the premises of the auctioneer, there were 14 auctioneers in India spread over the cardamom growing States, viz. 5 in Kerala, 2 in Tamil Nadu and 7 in Karnataka. Vandanmettu in Kerala, Pattiveeranpatty in Tamil Nadu and Sakalespur in Karnataka were the major auction centers in India. The auctioneer brings together the producers holding registration certificate and the dealers (also referred to as 'bidders') holding license issued by the Spices Board to deal in cardamom. As per the cardamom rules, the dealers shall not purchase cardamom from an estate owner who has not registered his estate or from an auctioneer who has not been licensed by the Spices Board (formerly Cardamom Board). Further, no dealer shall solicit or accept any amount in cash or in kind from cardamom producers or auctioneers whether by way of discount or commission.

A dealer participating in auction shall pay full value for the entire quantity of cardamom in the lot purchased by him. The cardamom exporters are another important link in the marketing system of cardamom. According to the Cardamom (Licensing and Marketing) Rules of 1977, the exporters shall not procure cardamom directly from a grower or from an auctioneer unless he is in possession of a license as dealer. The number of exporters varies from year to year as the license is issued for a year after which it should be renewed to carry on business. Cardamom market was characterized by high level of concentration at all levels. During 1980s, according to an earlier study, though there were 5 auctioneers in Kerala, one of them (CMC Vandamettu) accounted for more than 70 percent of the product sold through auctions. Apparently, the availability of large supply attracted large number of bidders and the consequent increased degree of competition pushes the price up. The realisation of higher price in the particular centre induces more sellers to register their product with that Centre. It was further observed that during 1971-85 only two auctioneers remained permanently in business and together they accounted for nearly 90 percent of the total sales in Kerala. It should also be noted that all these auction centres were located in the cardamom growing areas of Idukki District covering an area of 45,172 hectares in 18,585 holdings. The only exception was the auction centre at Cochin. In Waynad there was no auction centre though there are 270 holdings with an area of 4,247 hectares – a case of active exclusion according to the taxonomy of exclusion suggested by Amartya Sen. It is also to be noted that for historical reasons, larger number of cardamom growers are not having titledeeds and that they are not entitled to register their product with an auction centre – yet another case of active exclusion.

In 1983-84 there were 528 licensed dealers in India. It was found that while Kerala accounted for 69 percent of the total production, only 32 percent of the total dealers belonged to Kerala. There was a clear domination of Tamil Nadu in cardamom trade. That is, Tamil Nadu, with 6 percent of the total production accounted for 43 percent of the dealers. Historically, the activities of the traders were confined mainly two major markets of Tamil Nadu, viz. Bodinaikannur and Viruthnagar. The situation remains more or less the same at present. The dealers buy cardamom mainly at auctions and directly from growers. Even though all the traders are licensed to deal in cardamom marketing at every stage, all of them are not equally involved in trade. Only a few of them, about 125, used to deal in auction business. This could be because of the need for huge capital and risk associated with auction. Therefore, many of these dealers preferred to buy directly from growers. The dealers perform an important function of grading the product. They grade and pack it according to the demand from different consumers within the country and outside. With respect to concentration of market power, it was found that on an average 29 dealers bought cardamom from the major action centre (that accounted for over 60 per cent of the output sold in Kerala) during 1979-80 to 1983-84. Out of this, the share of top 5 dealers was found to be 58 percent in 1979-80 and it increased to 65 percent in

1983-84. Similarly, the share of top 2 dealers also increased during the same period. It was 34 percent in 1979-80 and it increased to 39 percent in 1983-84. This may be partly accounted for by the decline in the number of bidders participating in the auction. All the top 5 bidders are found to be exporters. Of these five, one dealer is found to be dominating in all the years (Joseph 1985).

At a time when cardamom was an export oriented crop5 the price released in the internal market depended to a great extent on the market strategy adopted by the exporters. This was more so, as the export trade was controlled by a few exporters. Export in 1976-77 was 8936 M.T and it increased to 2876 M.T in 1978-79 and showed a marked decline in 1982-83. With increase in the quantity exported the total number of exporters increased and as quantity exported declined the number of exporters also decreased. The interesting point to note is that while the average share of an exporter was nearly one percent, the top 10 exporters accounted for more than 60 percent of the total and some of these leading exporters are found to be the dominant buyers in certain important auction centres. Moreover, their share in the total export increased considerably during the period under consideration. Even though the export was subjected to yearly fluctuations, the share of top 5 exporters was reasonably stable at 50 percent in all the years under consideration (Joseph 1985). Thus the foregoing discussion reveals that prior to the introduction of e-auctions, the institutional innovations has had the effect of concentration of market power in the hands of a few at all levels. A few auctioneers accounted for bulk of the quantity sold through auctions and a few dealers accounted for bulk of the quantity bought through auctions. At the export level also a handful of exporters controlled most of the quantity exported.

Functioning of the Market

Having examined the marketing system, let us proceed to examine the functioning of the market which was governed by the institutional innovations in the form of marketing rules brought about by the erstwhile cardamom board. The growers register their produce with the auction centre, sufficiently before the auction, and may get an advance from the auctioneer which will be deducted from the sale proceeds. The auctioneer maintains a warehouse where the produce is kept till the time of auction. If the product is withdrawn from the auction, (the seller can withdraw the product if the price quoted is unsatisfactory) the seller is liable to pay the warehouse charges at the rate of 15 paise per kg per day. At the time of registering the produce for auction, the auctioneer mixes the produce thoroughly and takes a sample from each lot. The Cardamom (Licencing and Marketing) rules (1977) describes the procedure of taking the sample as follows: "The auctioneer shall draw 500 grammes out of each lot of cardamom offered for sale as sample. Expose 350 grammes out of it for bidders to examine at the place of auction four hours before the commencement of the auction and circulate a list indicating quantity of each lot, weight in grammes per litre7 and reserve price8. The 100 grammes out of the sample shall be given to the highest bidder and the balance of 50 grammes shall be kept by the auctioneer in a sealed polythene bag for a period of seven days for verification in case of any dispute. On expiry of the said period or on settlement of the dispute, as the case may be, the owner of the cardamom shall be entitled to receive back the cardamom from the auctioneer." But the actual procedure of taking sample and auctioning was stated to be different from what is stipulated by the rules. The auctioneers in Kerala take a sample of 750 grammes of which 500 grammes is paid at the rate at which the rest of the quantity is sold. Out of the 250 gramms, 100 grammes is given to the highest bidder and the rest 150 grammes is appropriated by the auctioneer in addition to the one per cent commission. Further, the auction report showed that 8 percent of the sales tax is taken from the seller as handling charges. Thus from each lot, whatever may be its size, 250 grammes is lost by the seller in addition to 8 percent of the sales tax as handling charges. The practice of deducting fixed quality from each lot as sample irrespective of the lot size, make the effective price (the price that the sellers get per kg after all deductions) much less than the actual price (the price quoted in auction). This discrepancy between actual price and effective price reduces as the lot size increases. Thus the registered sellers with a lot size below a certain minimum quantity may not find it profitable to sell their produce through the auction centres because of the institutionally set rules; a typical case of active exclusion that Amartya Sen (2000) described. Under the conventional auctions, it used to take place in the premises of the auctioneers and procedure used to be as follows. The dealers (bidders) will be seated around a round table and a small tray will be kept in front of each of them, and out of the 750 grammes of cardamom taken as sample from each lot, 500 grammes will be spread in all the trays for examining the lot. Then the intending buyers bid upwards for individual lots and the highest bidder will be given the lot. The process of bidding starts with the announcement of the lot number and lot size. It is a quick process and each lot takes hardly a minute to get auctioned. The seller has the freedom to withdraw her product from auction before the bidding is over. Once the price is determined through bidding, the seller loses his right over the produce. Once the bidding is completed, a slip showing the weight of the lot and the agreed price is given to the seller by the auctioneer, not the value of product sold. The slip also shows all the deductions and charges to be paid by the seller. The payment is made later, usually after 20 to 45 days as the case may be.

There is, thus, an in-built credit facility extended by the growers which in turn is guaranteed by the auctioneers. The possible explanation for this delay may be the huge capital requirement on the part of the dealers because of the high unit value of the product. Therefore, the growers will have to wait for the product to be exported or to be dispatched to the consuming centres. Some of the growers were of the view that the delay in payment has been perpetuated both by the dealers and the auctioneers. The auctioneer could refrain from paying to the sellers for the single reason that a few of the dealers have not made the payment even though all the others have paid. The delay in payment, therefore, keeps away those growers without the holding power (presumably the small holders) from the auction system a typical case of illusive exclusion.

A perusal of the auction procedure described above reveals that there are different types of exclusion inherent in the system resulting from the institutional innovations. There were a large number of growers who were not having the registration certificate and hence confronted with active exclusion and active exclusion also take place on account of the institutionally set practice of taking fixed sample from each lot regardless of the quantity offered for sale. From our discussion with the growers we came to know that they are forced to sell their produce at the auction centres, even at an unsatisfactory price, because of the inability to repay the advance already received by them, a typical case of illusive inclusion. If they withdraw the product they will have to pay the go down charges at the rate already mentioned. Thus the practice of giving advance at the time of registering the product for auction seems to have the effect of subordinated inclusion. It is pertinent to note that, the system of taking sample which prevailed in the auction centres in 1940s was on the basis of quantity sold. For example, in Bodinaikannur, a sample of 2 lbs was taken from lots up to 10 thulams (1 thulam=10 kg) and 3 lbs from the lots above 10 thulams and the payment was made at the time of sale. Therefore, the effective price realised was proportional to the price quoted the same for all sellers.

Price Variation across Different Lots under Conventional Auctions

Having examined the spaces of exclusion induced by the institutional innovations let us now examine the process of price formation in auctions and price variation across different size of lots sold through auctions to see if there are additional spaces exclusion being created. To begin with, let us be clear about the market power of different actors involved; from the sellers side there is the market power of the growers of cardamom, both the large and small, who could withdraw the product from auction if the price quoted in the auction is found to be unsatisfactory. The auctioneers, similar to the sellers, are also interested in obtaining higher prices for the following reasons. First and foremost, since the auctioneers were also the growers of cardamom, higher prices would lead to increase in their sale proceeds. Secondly, higher the price realised in auction, higher the commission accruing to them and thirdly, higher price quotation in a certain auction centre would attract more sellers to that centre thereby increasing the total revenue accruing to the auctioneers by way of taking sample. Coming to the market forces operating on the buyers' side it needs to be noted that exporters, as a category, consist of both large scale exporters and small scale exporters and the leading exporters are found to be the major buyers in the auction centres. Since the profit of the dealers and exporters depends on the margin between the export price and auction price, the dealers and exporters would try to depress the auction price with a view to increase their margin. This is because the export price has been exogenously determined and therefore the exporters may not be able to manipulate the export price. Hence, in the bidding process in the auction centres, the exporters quote the price in accordance with their price agreements with the importers to other countries. Under these conditions one could postulate the process of price formation in auction centres in the following way: In the bidding process the bidders (the exporters and dealers) would Endeavour to reduce the price, given the fact that they have already made commitment to their counterparts at a certain price, with a view to add to their margin. At the same time, the growers and the auctioneers would be interested in obtaining the maximum possible price. If the price quoted is unsatisfactory the sellers would withdraw the product from auction. The relevant question is, which category of sellers will be able to withdraw the product? Obviously, only those with adequate holding power and good storage facility could withdraw the product. On the other hand, those sellers without adequate holding power and who have already received an advance from the auctioneer will be forced to sell the product at the price quoted in the auction. Thus, there is the possibility of price discrimination across different lots sold through auctions wherein the smaller lots (smaller growers) realising a lower as compared to the larger lots. In this context, the pertinent question is to what extent the bidders could discriminate across different lots? This depends on the supply and demand conditions in the market. Given the export demand, if the domestic production is higher than what the export demand warrants (excess supply) the price quoted in the auction centres will be low and the price discrimination tends to be higher. On the contrary, in a situation of low production (excess demand) the price quoted in the auction centres would be relatively high and the price discrimination across different lots may be low.

Price Variation under e-auction

In a context wherein the traditional auction system has not been successful in ensuring remunerative prices for the small holders the Spices Board introduced e-auction in place of convention auctions with a view to make the marketing system more efficient and ensure remunerative prices to all growers by harnessing information technology. Under the e-auctions, major change is in the auctioning system and that there is hardly any change in the rules governing auctions or the way in which samples are taken or in the time of payment. Unlike in the earlier system, auctions do not take place in the premises of the auctioneer, instead in a state of the art auction centre established by the Spices Board with over 60 computer terminals networked to a server and a few large display boards that display the details regarding the lots being auctioned (like reserve price, lot number, quantity, number of bags current highest bid). In the new system, licensed dealers are provided with a user id and password. The dealers have to login to the system to participate in an auction. In each week a particular day is assigned for each auctioneer for auctioning the product registered with them.

The auction starts with the reserve price and a bid by a dealer is made with key depressions (each depression indicates an addition of fixed amount). As in the previous system the grower has the freedom to withdraw the product if the price is not satisfactory. Identity of bidders is protected during the auction process. Highest bidder's name is displayed only on the Auction Masters' terminal.

It should be noted that by the time e auction was introduced in 2010 in Kerala, significant changes had taken place in terms of total production, productivity and export intensity. There has been a significant increase in the production of cardamom from about 4400 MT in mid 1980s to over 10,000 MT in 2009-10. Similarly, productivity increased from about 55 kg per hectare in 1980-81 to 269 Kg per hectare in 2009-10. More importantly, share of export in production (export intensity) steadily declined from over 60 per cent in the mid 1980s to less than 10 per cent since 1990s. There have also been changes in the market structure. The domination of a single auctioneer prevailed in the mid 1980s is replaced today by an oligopolistic market structure .When it comes to market concentration at the level of dealers there appears to have been significant decline.

CHAPTER-V

Climate Change Adaptation Measures through Cardamom

Plant systems, and hence crop yield, are influenced by many climatic factors, and these factors, such as rainfall and temperature may act either synergistically or antagonistically with other factors in determining yield. The net change in yield is determined by the balance between negative and positive, direct and indirect effects that can affect production. Indirect effects arise from changes in the incidence and distribution of insect, pests and pathogens and soil properties. In general, climate variables directly affect the occurrence of insect and pest outbreaks and the frequency of these is therefore likely to change in the present and future. Therefore, the interplay of these factors along with the level of intensity of practices decides sustainability of agro ecosystems. Performance of crop plants in terms of growth, development, biomass accumulation and yields depends on acclimation ability to the climatic and environmental changes exercising specific tolerance mechanisms that involve a complex network of biochemical and molecular processes . Abiotic stresses such as extreme air temperature, drought and chemical toxicity represent serious limitations to agriculture causing significant yield reduction in major crop species . High temperature is known to disturb plant metabolism of most plants that survive high temperature through prevention and mitigation of overheating and due to the capacity of protoplasm to tolerate high temperature.

Globally, the Asian and Indian monsoons are very important for agricultural activity. Monsoon failure in one season can brought about severe yield reductions of crops in mountain agro ecosystems. The IPCC (2001) indicates that climate change is an inevitable process although the magnitude, timing, precise location and direction of these changes remain uncertain. The climate change in the tropics is often expressed to be less than in extra-tropical regions and this is not necessarily the case for e.g. larger than expected variations in the tropical atmospheric energy budget which have profound effect on tropical precipitation and surface temperature. In most tropical forest regions deforestation and degradation will still dominate over climate change effects for at least the medium term (decades) and therefore, the feedbacks between the land use change and local micro climate may play a critical role in determining the viability and sustainability of forest ecosystems . Climate change impacts food production systems, particularly in areas with large vulnerable populations as well as land cover land use change (LCLUC) can influence regional climate dynamics. Change in climatic variability and its impact on food production remains uncertain particularly in the tropics. Lobell et al. have reported that not all maize varieties had responded similarly to climate change. The yield of maize has been increased in cooler sites but significantly reduced in areas where the climate is hotter in Africa. Hence, Biophysical factors such as temperature and rainfall as well as relative humidity can strongly affect crop yields. Changes in seasons and its duration are an indicator, as well as an effect of climate change. They reflect variations that are occurring in the cycling of energy in the global environment and profoundly affect the balance of life in ecosystem, impact crop agriculture and water balance of a region. Since the duration of a season directly affects the occurrence of insect and pest outbreaks and the frequency of these is, therefore, likely to change in future which will force us to redesign our pest management strategies. Hence knowing the past changes in seasons helps us for future sustainability of the farming society. In this study seasons are defined based on meteorological variables, namely, rainfall and temperature, as they affect agriculture and the irrigation of the region, and also are commonly used in climate change impact studies. Species of cardamom are mainly grown in the tropical regions of Africa, Asia, South America and Central America. Among some twelve species of cardamom, aromatic cardamom, which is popularly called "Queen of spices" (*Elettaria cardamomum* Maton) is commercially dominant in the Indian cardamom hills, western hill ranges of Guatemala as well as Usumbara hills of Tanzania. In the Indian cardamom hills (ICH) local variety Njallani is widely cultivated covering nearly 80% of the cardamom reserve under cultivation. The remaining areas are also occupied by selections of other planters and cardamom research stations. The ICH is one of the global biodiversity hot spots where the majority of the tree species belong to C3 photosynthetic type, where as cardamom has C4 pathway. Most model studies of change in forest vegetation highlight the potential importance of temperature or drought thresholds that could lead to substantial forest decline in the future. Higher air temperature and drought can seriously affect cardamom yields up to 70% in non irrigated areas, therefore, higher summer temperature coupled with drought episodes are considered the strongest climatic limitations across most cardamom growing regions.

Cardamom is native to tropical rainforests of the Western Ghats of India at an altitude range of 600-1600 msl. In these ranges the air temperature show more seasonal variation with an annual average about 23º C. Rainfall is abundant and well distributed, ranging from 1800-4000 mm with a dry season lasting three to four months and the atmospheric humidity frequently approaching saturation during the main monsoon months along with dense frost. The optimum mean annual temperature range for cardamom is 18 - 23º C. Above 25ºC the development and ripening of cardamom capsules are accelerated, often leading to early maturity. Continuous exposure to high day time temperatures, as high as 32ºC, could lead to complete withering of leaves and young tillers. A relatively high air temperature during blossoming (April-May) especially if associated with a prolonged dry season may cause reduced pollination and abortion of cardamom flowers. Air humidity has a significant positive impact on vegetative and flowering phase of the cardamom. On the other hand, the decrease in relative humidity in combination with elevated day time temperature may limit the successful cultivation and higher yields of cardamom. Rainfall, an important climatic element associated with climate change affects cardamom productivity because cardamom is very sensitive to both excess rainfall and drought. Therefore rainfall prediction for future years and decades is of paramount importance in cardamom productivity and yield sustainability. Therefore, extreme temperatures, depending on their intensity, duration and speed of imposition can impair growth as well as of yields of cardamom. Combination of elevated air and soil temperatures found to be more deleterious for cardamom because along with yield reductions pest management becomes more intensive particularly for root grubs, nematodes and white flies Vast majority of cardamom plantations belong to small to medium holdings which makes the farming activity highly important in providing higher incomes and wealth distribution particularly in India and Guatemala. Cardamom under intensive management yields successfully even under nontraditional areas of ICH where the level of shade ranges from 40-50%. Copious rainfall with normal distribution is often responsible for good harvest in cardamom. Among the commodities of spices, small cardamom has been the second or the third costliest spices in the world. Except for individual climatologically station's work published literatures for cardamom on the present and future climate change are scarce and limited than it's nearby hill ecosystems like Palni hills and montane cloud forest systems in Tamil Nadu. Few studies have reported the change in season length which can influence some phenological changes in crop plants besides affecting water resources. Among climate variables, amount of rainfall and its distribution is the most influential one for cardamom than air temperature. Therefore this study has been very relevant to explain the change in climatic elements and implications for cardamom cultivation and sustainability.

Large Cardamom Farming in Changing Climatic and Socioeconomic Conditions in the Sikkim Himalayas

Large Cardamom Cultivars in Sikkim

Farmers grow as many as eight different local cultivars of cardamom, which have been developed and tested in different agroclimatic situations and under different farm management conditions. Of these, six varieties are widely grown at an altitude of 600–2,400 masl throughout Sikkim and in some parts of the Darjeeling hills.One of these, seremna, a variety developed by the Limboos of Hee-Bermiok in West Sikkim, is a location-specific cultivar that is tolerant to diseases. Seremna is the best-performing large cardamom cultivar in Sikkim. Another promising local cultivar is dzongu golsai, which was developed by the Lepchas of Dzongu. Although its cultivation in North Sikkim is limited (only in Timvong and Passingdong areas), it is considered the most disease-tolerant variety and is preferred by the majority of cardamom growers in these areas. The establishment of nurseries for the mass multiplication of dzongu golsai planting materials is slowly gaining momentum. Another variety, bharlang, is best suited for cultivation in middle to high altitudes (1,000–1,800 masl) and has a high market value due to its comparatively high productivity, large capsule size, aroma, and characteristic maroon colour. During early 2000, the Indian Cardamom Research Institute (ICRI) of the Spices Board in Gangtok identified a potential disease tolerant variety of cardamom at the farmers' field and through selective breeding developed the cultivars 'Sikkim I' and 'Sikkim II'. The total six accessions collected from Sikkim I and II were planted in a germplasm conservatory. The plants were then transferred to multi-location trial plots and evaluated for comparative yield assessment. Both the varieties were identified as potential improved varieties and the genetic material was released by the Central Varietal Release Committee in 2002.

Trends in Large Cardamom Farming

Sikkim produces about 40% of the world's large cardamom, standing second after Nepal. However, since 2006, more than 60% of the plantations have become non-productive resulting in a tremendous decline in cultivated area. The income of marginal and cardamom dependent farmers has also dramatically declined, which has led to a marked increase in the number of farmers shifting to other income generating activities such as other cash crops, the tourism industry, government jobs, and so forth. As per the records of the Spices Board and Horticulture and Cash Crop Development Department of Sikkim, the total area under large cardamom production between 1999 and 2001 was 23,484 hectares and the actual production area was 19,912 hectares. In 2003, this area increased to 26,734 hectares, as a result of the area expansion mission initiated by the Food Security and Agriculture Department and the Horticulture and Cash Crop Development Department, Government of Sikkim. Accordingly, the actual cardamom yielding area increased to 22,714 hectares in 2003. In the following years, as a consequence of long dry spells and disease infestations, the area under large cardamom decreased, and by 2007/08 it was 12,500 hectares, which showed a decline of 45% In an effort to address this issue, government agencies and the Spices Board, under its area expansion mission, initiated a nursery programme to produce planting material for affected cardamom growers. This slightly increased the area under cultivation to 14,520 hectares in 2010.

The quantity of cardamom produced has followed a similar trend since 1999. In 2004, Sikkim saw record production of cardamom (5,152 metric tones increasing from 3,710 metric tonnes in 1999) and became the largest producer in the world . However, after 2004, there was a continual decline in production. By 2008, production had declined to 2,745 metric tonnes, bringing Sikkim down to second position after Nepal.

However, from 2010 to 2012 production gained a little momentum, owing to increased awareness, motivation among farmers, and the provision of extension services by agencies aiming to raise the production area (HCCDD 2011). As a result, the total production of large cardamom increased by around 14% from that of 2007/08 to 3,316 metric tones fetching around USD 47.54 million. Similarly, the yield also decreased to the bare minimum (148 kg/ha) during early 1990s, after which it started improving and, in 2010, reached 234 kg/ha. The average yielding area of large cardamom over the last 11 years (1999–2010) was 18,544 hectares.

During this period, the average production of large cardamom was 3,875 metric tonnes per year with an average yield of 207 kg/ha. Linear regression analysis showed a negative correlation between cardamom plantation area (hectares) and production (metric tonnes) over time. This means that area under large cardamom plantation as well as its production has declined over the years. Cardamom production and plantation area show a positive linear relationship, which means that there is a positive correlation between its production and area under production.

Currently, 16,037 of the total 111,830 households in Sikkim (14.34%) have a large cardamom plantation on their farms. Among the various social groups in Sikkim, Scheduled Tribes posses around 23% of the cardamom farms, Most Backward Castes hold 13% and Other Backward Castes hold 11%. Sikkim's East District has the highest number of households (5,275) with large cardamom farms followed by the West (5,060), South (2,937), and North (2,759) districts. Of the 16,037 households in Sikkim with large cardamom farms, the total marginal households with less than 1 hectare of land under cardamom plantation was 14,476, of which 10,975 households had less than 0.5 hectares and the remaining 3,501 had between 0.5–1.0 hectare under cardamom. Around 1,060 households possessed 1–2 hectares of cardamom plantation, 143 possessed 2–4 hectares, and 321 possessed 4–10 hectares of land. As many as 37 households possessed more than 10 hectares of land.

Climate Change and Global Warming

The plantation sector in the country is located mostly in the ecologically fragile locations. Therefore, there is a two way link between the ecology of these regions and the economy of plantations. There are indications to suggest that the productivity of plantation crops is influenced by the state of environment (Nair et al 1989) and hence the climate change. Studies have also shown that the coping up strategies of growers, like felling of shade trees for sale in coffee plantations for sustaining the revenue flows in the event of decline in prices, are having deleterious consequences on the environment (Damodarn 2002). Similarly, in cardamom plantations, felling of trees for firewood for cardamom curing is also having similar outcomes. This in turn has induced the Spices Board to replace the traditional fire wood based curing houses with modern curing facilities using electricity and/or gas. The adverse impact of excessive use of chemical fertilizers and insecticides and pesticides in plantations on the environment and health hazards' also has been a matter of concern. This could have adverse effect in the export market in a context of growing concern on environment and human/animal health. The unsustainable cultural practices and the consequent environmental degradation also adversely affected the land productivity. This has induced the commodity Boards to undertake new measures to make plantation production more environmentally friendly like, organic farming. Nonetheless the nexus between ecology and economy in the context of plantations is an area where we are confronted with number issues on which our understanding at best rudimentary and calls for further research to facilitate informed policy making and institutional interventions.

Climate Change and the Kyoto Protocol

Scientific evidence indicates that global climate change has been occurring for the past several decades and increasing greenhouse gas emissions will continue to cause climatic changes such as higher seasonal temperatures and increase the risk of extreme weather events. As a result of excessive use of pesticides and chemicals along with other environmentally unfriendly practices the agricultural sector also contributes to the GHG emission.

The Kyoto Protocol requires the rich and industrialized countries of the world to reduce their collective carbon dioxide emission to at least 5.2% below their 1990 emissions levels between 2008-2012. Domestic actions required to meet the Kyoto compliance targets can be very expensive for the developed countries- both financially and politically. Therefore, the Kyoto Protocol established three market mechanisms .

Carbon Credit: Opportunity for Natural Rubber and Cardamom

- The use of biomass gassifier, biogas production from latex processing effluents, potential use of natural rubber seed oil as a bio-diesel are strong cases that can attract CDM funding in the NR sector. CDM Executive Board has already approved modalities and procedures for such CDM project.
- Rubberized roads have been known to improve the fuel efficiency to transport fleet, which is eligible for CDM funding under energy use efficiency. The additional cost that has to be incurred in the rubberization of roads could be met from the potential CDM cash flow into the project.
- The possibility of cultivating natural rubber as part of reforestation activity in a degraded region with the exclusive aim of supplying the rubber produced from there into the market where it will directly substitute synthetic rubber. Such a project would generate CERs as a result of the carbon sequestration by the plants. In addition, the carbon dioxide equivalent of the amount of synthetic rubber displaced and the carbon dioxide equivalent of the energy that was required to synthesize that much amount of synthetic rubber will earn clean and strong CERs in the CDM market. Both in terms of the sale of the CERs and the rubber, such a project will ensure a captive and pre-determined market with due financial security from the start. In the fast changing global economic scenario, financing natural rubber cultivation from internally generated cash flows such as from CDM as discussed above needs to be explored on a priority basis
- In the case of Cardamom, in a situation of declining area under cultivation and an urgent need to increase area, the strategy to bringing new areas by planting shade trees will be eligible for CERs as a result of the carbon sequestration by the plants. Here it is worth noting that such initiatives are already being undertaken in Cambodia at the instance of Civil Society Organizations and Government.

Impact of Climate Change on Large Cardamom Farming

Climate change is a well-known phenomenon in the Himalayas and is causing unpredictable and erratic weather conditions, the rapid melting of snow, and the drying up of water sources (Sharma et al. 2009a ; Chaudhary and Bawa 2011; Chaudhary et al. 2011). Climate has a variety of affects on mountain ecosystems and the livelihoods of mountain people including on agriculture, biodiversity, water availability, and water quality (Chaudhary et al. 2011, Chettri et al. 2012): Local communities in the study areas appear to have substantial awareness of climate change and their observations concur with the scientific research on climate change trends, the flowering cycle of various crops and plants, and agricultural productivity (Chaudhary et al. 2011). An article published by Chaudhary and Bawa (2011) in Biology Letters validated local perceptions of climate change and its concurrence with scientific evidence in the Himalayas. Climate change has had a pronounced impact on large cardamom-based traditional farming, as well as on pollinator species, especially honeybees and bumblebees. Earlier studies in the western Himalayas conducted by Partap and Partap (1997, 2002, 2009) and Partap and Tang (2012) found that climate change-induced changes in local weather conditions affect the pollination of various crops by affecting the plant-pollinator relationship. Climate change affects the emergence and foraging activities of pollinators and influences the flowering time and duration of various crops and plants (Partap and Partap 2002; Partap and Tang 2012). However, so far, farmers in Sikkim have not experienced pollination-related productivity issues in large cardamom.

Climate change-induced changes in local weather patterns are reported to have increased the incidence of diseases and pests in various crops. Studies conducted by Partap and Partap (2002, 2009) in the western Himalayan state of Himachal Pradesh and by Partap and Tang (2012) in Maoxian County in the southwestern part of Sichuan province of China reported similar findings in apple and other fruit crops.A few studies have reported the impacts of climate change in the Sikkim Himalayan region. Rahman et al. (2012)have reported that the mean minimum temperature has increased by 1.95°C while maximum temperature did notexhibit any significant departure from long term average. They further reported that rainfall over 30 year periodincreased by 124 mm (measured from 1981 to 2010 at Tadong,1350 m). The communities in the region have enough awareness to show climate change trends, phenology and agriculture productivity in the region which are in concurrence with scientific research (Chettri et al 2012). Sharma (2012) reported that the phonology of a number of multipurpose agroforestry species in the cultivated systems of Sikkim Himalaya has altered over the last 10 years. Sharma and Rai (2012) have described pronounced effect on agricultural crops in low, medium and high altitudes, and also in response to reduced pasture lands in the greater and trans-Himalayan region. They also emphasized traditional ecological knowledge, agrobiodiversity, land uses and diversification of livelihoods among local communities to cope with changes, and that each of these will play an important role in true adaptation to climate change. Sharma et al. (2010) and Tambe et al. (2012) have reported indirect evidence of decline of spring water availability during the dry season and have suggested an approach to monitor spring water flows, which will not only address the needs for irrigation facilities in agriculture but also help provide drinking water to rural communities. Sikkim has completed micro-level climate change related vulnerability assessment of the rural communities at the Gram Panchayat level. The state Action Plan on Climate Change has been prepared after detailed multi-sectoral consultations and village participatory rural appraisals in consonance with the National Action Plan on Climate of the Government of India (Arrawatia and Tambe 2012). In the Khanchendzonga Landscape, the ICIMOD has been promoting large scale conservation planning for promoting habitat connectivity through corridors development, as well as building capacity of the community for economic interventions based on conservation linked livelihood options (Chettri et al 2008, 2012).

Farmers' Perceptions and Adaptation Practices

Farmers mostly from the mid and lower altitudes, felt that the weather during the onset of cardamom flowering season was becoming drier, while 4.5% (mostly from the high altitudes) felt it was becoming wetter. The rest of them perceived no change or fluctuating weather conditions during cardamom flowering season. Besides, during the peak flowering season for cardamom (May–June), erratic rainfall events caused flowers to fall and decay, resulting in the loss of fruiting intensity during September– October. Almost all of the cardamom growers and key informants perceived the climatic conditions to be changing and said that rising temperatures have led them to harvest cardamom 10–15 days earlier. Such changes also affect the yield and quality of the produce. Around 66% of the respondents observed that snowfall, a yearly phenomenon 15–20 years ago, has become rare during the past 4–5 years. Thus, there is a need to conduct research for the development of varieties suitable for cultivation under changing climate conditions at various altitudinal locations.

Interventions Required for the Revival of Large Cardamom Farming

Large cardamom continues to be the lead crop in the farm economy of Sikkim and, therefore, warrants immediate intervention for its revival. This came up strongly in discussions with farmers who identified several interventions required for the revival of this crop. There is an increasing realization that the problems associated with large cardamom farming can be substantially reduced by adopting appropriate measures to maximize the economic and ecological functions of cardamom farming. In fact, related institutions are making efforts to revive large cardamom, particularly in areas that have the most potential for cardamom production. Major revival practices include the provision of irrigation facilities, soil fertility maintenance strategies, the plantation or re-plantation of disease resistant or disease tolerant varieties, awareness and training on cardamom farming, and financial support and subsidies. These revival practices are now widely practiced in Hee-Pechreak and Hee-Martam in West Sikkim, but are yet to reach other more remote cardamom growing areas in Sikkim.

Climate Change and its Implications for Cardamom Farming and Sustainability

The strong variation in the air temperature of the investigated ICH region is remarkable compared to the global scale warming. Brohan et al. reported a warming trend of 0.33ºC/decade for the period 1979-2005 on land in the northern hemisphere and of 0.27ºC/decade in both hemispheres. The steady increase is noticed in the mean annual temperature range for Pampadumpara by about 0.1ºC per year during the period 1978-2007. The reason for such abrupt increase in temperature climatology for highly disturbed cardamom agro forestry system could be attributed to the lee warming and high velocity direct dry winds experienced by the ICH plateau. The other two stations namely; Myladumpara and Vandiperiyar, later lie interior in the deep valleys of Periyar Hills than the plateau region of ICH, has experienced decreasing trends of annual mean temperature range (0.16 and 0.015ºC per year during the period 1989-2007). Jauregui [30] had reported similar warming trends for other places in related tropical regions where the increase in mean annual temperature range was 0.12ºC/year (1956-1983). Earlier studies showed an increase of air temperature by 0.3ºC for ICH Murugan and Miniraj. The yields of most of our food, feed and fiber crops decline precipitously at temperatures much above 30ºC [32]. Cardamom, under rain fed condition, can tolerate air temperatures up to 32.2ºC, and beyond this level both growth and development as well as yield is reduced. Analysis of monthly temperature regime since 2000 showed temperatures more than 33.5ºC during the summer Leaf scorching, reduced growth and development and flowering and increased leaf spot diseases are the common effects of severe warming in ICH. Because of the rise in air temperature coupled with regular thinning of forest trees the soil temperature has increased significantly at least in the top 10 cm. The increased frequency of break periods during monsoon months, with elevated soil temperature, favored soil insect pests like root grubs and plant parasitic nematodes besides white flies and thrips. The root grubs and nematode infestation on cardamom was less severe at least up to 1995 and were considered minor ones. In 1992 and 1993, outbreak of white flies was reported in Kumuli and Mali hot spots of cardamom as of weather change that was conducive for their development. Since then planters have been using lot of pesticides every season for controlling white flies. These pests have become major pests now because of soil-climatic change.(Murugan et al 2012)

The increase in atmospheric temperature in ICH is significant from monthly to five yearly scales. Such an increase in the temperature causes the diurnal temperature range to be more negative. The general decrease in precipitation level especially for the monsoon months can affect cardamom productivity as it is sensitive both to drought and excess rainfall. Large uncertainties exists in future rainfall predictions based on GCM has to be considered and interpreted carefully that too in tropical mountain ecosystems like ICH. Significant variability in precipitation patterns both in valleys and peaks of the ICH can also influence yield productivity of cardamom. The increase in atmospheric temperature is asymmetric as the mean minimum temperature increases faster than the mean maximum temperature. Diurnal temperature variation has more influence on cardamom phenology and productivity. Significant positive response of cardamom yield to summer rainfall and higher mean minimum temperature as well as enhanced relative humidity shows that the currently popular variety Njallani could be a good choice for ICH. Along with atmospheric temperature, the soil temperature is also expected to increase causing enhanced population of root feeding soil insect pests. The population dynamics and distribution, as well as the severity or intensity of insect pests and disease-causing pathogens would vary according to the rate of climate change in the ICH, therefore, the current pattern of pesticide use on cardamom and tea would change accordingly, irrespective of the variety in use. As cardamom farming in ICH continues to modernize and intensify, concerns about the loading of plant nutrients and pesticides on the ecosystem will likely to increase. The pathways of both nutrients and pesticides applied can be extremely complicated in such mountainous agro forest ecosystem, so continued research effort should be placed on better understanding of nutrient dynamics and pesticide degradations to address both agronomic and environmental aspect of the ecosystem. Managing cardamom ecosystem, especially forest and water quality protection and profitable cardamom farming could be the major future challenges.

The Importance of Large Cardamom to Livelihoods

The agricultural economy in the Sikkim Himalayas is largely based on high-value cash crops, with large cardamom being the most important cash crop. Rural farming families have by and large built their lifestyle around the income generated from large cardamom in terms of food, housing, education, health, social activities, and farm management. The decline in cardamom production has had a direct impact on local livelihoods. Large cardamom is the second largest contributor to household income in the study areas after employment and remittances, fetching around USD 910 per family per year. This study estimated that almost one-third of household income is derived from large cardamom. Earlier studies by Gupta et al.

(1984), Sharma and Sharma (1997), Sharma et al. (2000), Avasthe et al. (2011), and Singh et al. (2011) made similar estimates of the contribution of large cardamom to household income. The study by Sharma et al. (2000) further reported that the gross income of households was almost double in large cardamom-dominated systems, compared to traditional mixed-forest systems.

However, the contribution of large cardamom to overall household cash income has declined from about 50% in 1997 (Sharma and Sharma 1997) to 38% in 2000 (Sharma et al. 2000) and 29% in 2013 (Sharma 2013). There are two possible explanations for this: Firstly, the emergence of new income generating opportunities such as government jobs, work under the MGNREGA, and employment in tourism sector, as reported by Sharma (2013), and, secondly, a decline in cardamom farming area and production due to the emergence of new diseases and pests as a result of climate change-induced changes in local weather conditions (Sharma 2013). Yet, large cardamom continues to be the main cash crop in Sikkim and second most important source of household cash income in the study areas.

The trees in large cardamom-based agroforestry systems provide fodder, timber, and fuelwood and are important livelihood sources for rural households. Large cardamom farming is thus a self-sustaining system and has been adapted independently by low-income, marginal farmers of Sikkim. A number of institutions in Sikkim and Darjeeling are currently working to improve rural livelihoods through research and development, the provision of subsidies, technical expertise, and capacity building. In Sikkim, the Horticulture and Cash Crops Development Department, and the Food Security and Agriculture Department of the Government of Sikkim are collaborating with the Indian Cardamom Research Institute (ICRI) of Spices Board of India to provide expert extension services for improving large cardamom farming (FSAD 2010; HCCDD 2012; Sharma and Acharya 2013; Spices Board 2004, 2014). Different departments of the Government of Sikkim and the Spices Board have initiated a number of programme activities including the development of quality planting materials, establishment of nurseries under a certified nursery scheme, a replanting scheme, the supply of sprinkler irrigation units, and low-cost cardamom curing plants over the last 30 years. However, some of these schemes are not yet fully acceptble to cardamom growers in local situations. In most cases these schemes use a linear and sectoral approach of training farmers on new techniques and disseminating scientific findings for best management practices. Unfortunately, there is a lack of interdepartmental convergence when it comes to the implementation of 'improvement' schemes. Moreover, though the beneficiaries of these schemes are always the farmers, no effortis made to include them in the process of designing these interventions. The loss of large cardamom cultivation practices brings with it a host of interconnected issues, including a change in land use patterns, a loss of genetic biodiversity, family members seeking income from external sources, rural to urban migration of youth, and ultimately the loss of traditional knowledge systems in Sikkim (Hunsdorfer 2013).

Pollination of Large Cardamom: The Role of Apis Cerana

With regards to pollination, large cardamom is a self-fertile crop, but requires cross-pollination to produce fruit/capsule. However, literature is scarce on the pollination and pollinators of this crop. There have only been a few research studies carried out on large cardamom pollination. Bumblebees including *Bombus braviceps* Smith and Bombus *haemorrhoidalis* are reported to be its most efficient pollinators (Deka et al. 2011; Kishore et al. 2011; Sinu and Shivanna 2007; Sinu et al. 2011). Singh et al (2011) reported several species of pollinators including bees, butterflies, beetles, and mammals as important pollinators of various crops cultivated in Sikkim. These authors further reported the declining number of hive-kept honeybees and wild colonies of honeybees as one of the reasons for declining crop production.

A recent study by Sinu et al. (2011) reported *Bombus haemorrhoidalis, Apis cerana, Megachile lanata, Episyrphus ballatus* (hover fly), *Macroglossum stellatarum* (howk moth), and *Aethopyga siparaja* (crimson sunbird) as common visitors of cardamom flowers. Among these *Bombus haemorrhoidalis, Megachile lanata,* and *Aethopyga siparaja* were found to be the most efficient pollinators based on the transfer of pollen to stigma following a single visit of the pollinator to virgin stigma.

In relation to the indigenous honeybee, studies by Sinu and Shivanna (2007), Deka et al. (2011), and Kishore et al. (2011) concluded that *Apis cerana* is a pollen robber. Sinu and Shivanna (2007) reported that *Apis cerana* bees visit cardamom flowers to collect pollen, but do not touch the stigma and, hence, do not pollinate the flower. They reported that *Apis cerana* leave 70% of large cardamom flowers un-pollinated, while collecting all the pollen from the flower, thus negatively affecting its pollination chances (even by other visitors) resulting in severe crop declines.

However, farmers in this study reported they have not noticed such a decline in their large cardamom yield as a result of *Apis cerana* visits, even though *Apis cerana* beekeeping is common in the cardamom farming areas of Sikkim. A possible explanation for this is that the previous studies were conducted in a controlled environment where no other pollinators were allowed to visit the flowers. Under open field conditions *Apis cerana* is not the only insect visiting cardamom flowers, but one among many including different species of bumblebees. In-depth research on large cardamom pollination requirements, its pollinators, and their effectiveness is needed to validate these findings under open field conditions.

Recommended Interventions

Urgent research and development interventions are required to establish disease free varieties of large cardamom; improve farm management; and conserve effective pollinator species to counter the impacts of climate change and the decline in production and yield. Interventions are also needed to improve post-harvest technology to improve quality, thereby increasing the price received by farmers.

Large cardamom farming is less labour intensive than any other farming system in Sikkim and does not require large or expensive infrastructure at the growers' level. However, a growers association could give farmers greater negotiating power, enabling them to negotiate fair prices for their produce. The North Eastern Regional Agricultural Marketing Corporation has developed a marketing channel whereby auctions are held monthly in Rangpo, a nearby market town in Sikkim, to assist growers to sell their produce at better prices. However, growers' cooperatives that could pool produce and bring it to primary auctioning centres are still lacking.

Apart from the management and upkeep required by large cardamom, some future objectives could be to further improve the quality of planting materials through certified sapling nurseries, increase the spread of best management practices, facilitate more efficient farmer innovation processes through communication, improve product quality through improved post-harvest technology, and increase market channel efficiency.

CHAPTER-VI

Recommendations on Structural Infirmities in India's Plantation Sector

Short-term, Medium-term Term and Long Term Measures

Plantation sector is not just a foreign exchange earner but a Key Sector in India's Inclusive Growth Strategy because of its role in employment generation, (especially women labour) livelihood of small and marginal growers, backward communities, balanced regional development and addressing environmental issues. Further it is, perhaps, the only sector where competitiveness determines decisively the ability to foster inclusive growth as the competing countries are "desperate exporters" with hardly any domestic market. However, the sector is confronted with series of structural rigidities that creates spaces of exclusion at different levels.

Plantation Sector as Whole

The long term vision shall be to build a system of production and innovation that is competitive and inclusive through modernising plantation sector. This will involve a transition from factor intensive to knowledge intensive production system through investment in plantation infrastructure (physical, human & knowledge) and initiating measures to ensure stable and remunerative prices. Short term/ medium term measures to achieve the above include;

Subsidy and Cess

- Since Subsidy for replanting is paid ex post it shall aim at covering the cost of interest on investment not the capital component per se and shall be paid directly to the growers' bank account.
- Subsidy shall have a base component on the basis of effort and a supplementary subsidy based on performance.
- Cess on commodities shall be a percentage of the price instead of a fixed amount.

Labour, Employment and Social Cost

- Adaptation of NREGP in plantation regions: plantation work shall be treated on par with creation of social assets: Not to coincide with peak work seasons .
- Plantation sector shall be made attractive to the new generation through measures like using new term "plantation curator" (or any other term that highlight their role in

environmental protection) instead of labour: devising a decent uniform; use more machines, provision of housing colonies.

- The social provisioning for labour shall be the responsibility of the State with contribution from private sector.
- Encourage mechanization, formation of labour banks *ala* Rubber Board and introduce productivity linked reward system.

Prices

- Strengthen and Reform Price Stabilization Fund with at least Rs 10,000 Crores as the corpus given the important role of plantation sector in inclusive growth (It will ensure a payment of at least Rs 20,000 per ha if the price falls below a threshold level).

Physical, Human and Knowledge Infrastructure

- Establish Plantation Modernization Fund of Rs 5000 Crores for improving the Physical, human and knowledge infrastructure.
- Scale up R&D activities of the institutes under different commodity boards and facilitate greater interaction between them to learn from mutual experience.
- Formation of farmer-researcher groups for harnessing traditional knowledge systems for conservation, development and use of local and traditional varieties and development of new processing devices.
- Commodity boards shall emerge as source of ideas not objects by harnessing the S&T system & knowledge base in the country. Let their focus be on applied research and basic research to be undertaken by the universities and national research institutes.

  Harness the Southern RTAs and FTAs to address the commodity problem through South -South cooperation & knowledge sharing ala the EMBRAPA of Brazil. (The ASEAN India vision 2020 has already incorporated this).
- Ministry of Commerce shall consider setting up An Empowered Committee of State Ministers concerned to address the issues involving Central and State Governments

Recommendations: Spices

Organizational Measures

- Reform of Board membership to represent more crops and small holders
- Benefits provided for cardamom growers shall be extended to all the other crops covered under the Spices Act.

- Restructuring of the Spices Board to form divisions/departments for each/group of spices by increasing its manpower base.
- Formation farmers' groups to be active participants in all the field level activities of the Board.

Production & Environment

- Certification for bio fertilizers and pesticides.
- Karnataka-specific package of practices and intensifying the extension services in Karnataka.
- Bring new areas by planting new trees for shade, preferably the ones that bring additional income like jack fruit, nutmeg, mangostein, Rambutan. And explore the possibilities for claiming carbon credit.

Marketing and export

- Introduction of e-auction has given rich dividends; remaining problems call for Amendment of Cardamom (marketing) rules, 1977 such that;
- Spot payment shall be made in cardamom auctions; if not the delay in the payment for cardamom sold through auctions be reduced at least to seven days.
- For emerging as a global hub for processing; extend the export oriented production measures and the spice park facility to all the spice crops at decentralized level and invest heavily in brand building.
- In light of export cess being done away with the Board shall consider introducing a production cess ala Rubber Board.

R&D and Labour Welfare

- Extend the R&D agenda to include other spices and scale up the R&D activities.
- Labour welfare schemes by the Board may be expanded ala Rubber Board.

Recommendations for Natural Rubber

Production

- Consider provision of subsidy on the basis of the number of trees planted instead of the present system of subsidy per hectare.
- Increase the amount of replanting subsidy as it compares poorly with the competing countries.
- The Rubber Board shall consider certification for the planting materials.

- The Board shall consider issuing ID cards/smart cards for the growers ala Spices Board.
- Rubber Board shall explore the biotechnology (or other) options for developing a pulse variety (or any other crop) that grows in rubber plantations and bring additional revenue for the farmers.
- Formation of Rubber Producers' Society (RPS) is a very useful experiment. But today it covers only 20 per cent of the growers and hence calls for expanding its coverage to include all growers.

Market and Price

- Marketing of both sheets and latex needs to be made more efficient.
- Consider introducing more scientific methods for deciding quality and price of sheet rubber.

Environment

Strengthen and support the Board's initiatives to use biomass gassifier, biogas production from latex processing effluents, use of natural rubber seed oil as a bio-diesel, rubberised roads that are eligible for CDM funding in the NR sector.

Development Regimes: Subsidized Replanting, Rejuvenation and Other Schemes

There is a gestation lag between planting and harvesting for both cardamom and natural rubber, and the yield cycle involves broadly four phases; though the duration of each phase vary from one crop to another. In the case of natural rubber, there is an initial pre-bearing phase of about seven years followed by an early harvesting phase of about one to three years wherein yield is positive and increasing with high variability. During the third phase (may be called peak bearing phase) of about four to 13 years yield reaches the highest level and in the last phase the yield declines. Since the age of the plant has a crucial bearing (along with other factors) on the yield level, timely replanting is needed to maintain the age profile of the plantations in such a way that the share of old-aged plants is minimized. Along with technical advice and training and other farm inputs, both Rubber Board and Spices Board have been providing financial support for replanting and new planting to induce the growers to undertake timely replanting and to bring in new area under cultivation. Both commodity Boards are also providing assistance for setting up facilities for post harvest and processing activities.

Cardamom (small)-Replanting and Rejuvenation

Scheme

The basic premise behind this scheme is that in order to maintain and sustain high levels of productivity, it is important that old cardamom estates are replanted at the right time. It is reported that one of the important reasons for the high productivity in Guatemala is the practice of regular replanting when the plants attain the age of about seven to eight years.[1] The systematic replanting efforts ensure that their plantations are always juvenile, healthy and highly productive. In India also, there are evidences to the effect that the planters who adopt timely replanting and follow scientific cultivation practices realize yield levels much higher than the Guatemala yield. While about 10 per cent of the total area in India needs to be replanted every year, actual area replanted annually is much less. Since 95 per cent of the growers in India belong to the small and marginal category, they are reluctant to remove the uneconomic plants and wait for the new plants to achieve maturity.

In the case of large cardamom also it is observed that lack of technical know-how, non availability of quality planting materials, presence of senile, old and uneconomic plants are the major factors affecting production and productivity. Hence, along with technical support, financial assistance in the form of subsidy is provided by the Spices Board to induce framers to undertake replanting. The replanting programme, therefore, is intended to encourage registered small and marginal growers of cardamom to take up replantation of old, senile and uneconomic plantations of small cardamom in Kerala ,Tamil Nadu and Karnataka and large cardamom, in Sikkim and Darjeeling district of West Bengal using disease free and healthy planting materials.

In the case of small cardamom, as of now subsidy for replanting amounts to Rs.39,171 per ha and Rs.29,675 per ha respectively for the small (up to 4 ha) and marginal growers (4 to 8 ha). The subsidy covers 33% and 25% of the cost of replanting and maintenance during gestation period in the States of Kerala and Tamil Nadu. In Karnataka, the per ha. subsidy offered is Rs.29,919/- for holdings up to 4 ha. and Rs.22,666/- for holdings above 4 and up to 8 ha. being 33% and 25% of the cost of replanting and maintenance during gestation period. In case of large cardamom, a per hectare subsidy of Rs.16,500/- and Rs.12,500/- is offered to small (up 4 ha) and marginal growers (4 to 8 ha) growers. As in the case of small cardamom, it accounts for about 33 per cent and 25 per cent of the replanting cost per hectare. (During 2008-09, an area of 850 ha. was brought under replanting at a total expenditure of Rs.1.048 crores.)

Rejuvenation

Under rejuvenation programme, poor yielding plants in the existing plantations are identified and removed and the gaps thus formed are filled with quality planting material. In addition to this gap filling, scientific plant protection operations, fertilizer application, inter-culture operations, irrigation and other good agricultural practices as per the recommended package of practices are adopted. This programme is implemented in the States of Kerala and Tamil Nadu for registered small and marginal growers of cardamom with a holding size up to 4 hectares. The subsidy provided for rejuvenation is Rs.14025/- per hectare. In the case of large cardamom the subsidy offered is Rs.6600/- per hectare for holding size up to 4 hectares.

Certified Nursery Scheme

This scheme was started as a complement to the replanting scheme. With the introduction of the replanting scheme there has been a substantial increase in the demand for seedlings. On an average, it has been estimated that about 50 lakhs cardamom seedlings were required for the planting purpose. The Board's departmental nurseries have an annual production target of about 5.5 lakh seedlings. The annual contribution of private agencies was estimated at about 15 lakh seedlings. There was a gap of about 29.5 lakh seedlings.

In order to produce disease free, healthy and quality planting materials, certified nurseries are opened in growers' field under the technical supervision/ guidance of Spices Board. In Karnataka, the planting materials are produced through bed nurseries and poly bag nurseries whereas in Kerala it is produced through sucker multiplication nurseries. During 2008-09 the subsidy offered was Rs.1.25 per planting material in Karnataka and Rs.1.75 per planting material in Kerala and Tamil Nadu regions.

In the case of large cardamom, the Board supports raising of sucker nurseries in farmers' field by offering a subsidy of Rs.1.15 per sucker. During the 2008-09 19.98 lakhs cardamom suckers were produced from the certified nurseries opened during the previous season in growers field.

Irrigation and Land Development

The programme aims to provide water resources in cardamom plantations by constructing water storage devices like farm ponds and wells. Installation of irrigation equipments, soil & water conservation works are also supported. The Board is implementing the programme in Kerala, Tamil Nadu and Karnataka. In Karnataka the programme is implemented jointly by the Spices Board and State Government under their Western Ghat Development Programme, sharing the cost on a 60:40 basis by the State and the Spices Board. In Kerala and Tamil Nadu, full amount of subsidy is paid by the Board due to non availability of funds from the respective State Governments. The programme provides financial support to farmers by way of subsidy ranging from 25 to 50% of the unit cost approved by NABARD.

During 2008-09, in Kerala, a total number of 625 water storage devices were constructed, 148 irrigation equipments installed and 12.18 ha. were covered under soil conservation. In Tamil Nadu, 7 water harvesting devices were constructed and 4 irrigation equipments were installed under the programme. In Karnataka, 78 devices were constructed and 60 irrigation equipments were installed during the season. The total expenditure towards payment of subsidy under the programme was Rs.1.127 crores, covering an area of 1000 ha.

Rain Water Harvesting Devices

Irrigation during summer months is essential in Cardamom plantations for getting a higher yield. A cheap method of harvesting rain water for irrigation purpose in the Cardamom plantations is using excavated storage tanks lined with UV resistant polythene tarpaulin also called Silpauline. This method of harvesting rain water for irrigation purpose is adopted by cardamom growers because of its low cost and convenience. Spices Board is popularizing this method for irrigating cardamom plantations in Kerala, Karnataka and TamilNadu. It is estimated that a storage tank of 200 cu. metre capacity (eg. 16m x 5 m x 2.5m lined with Silpauline) can store about two lakh litres of rain water, which is sufficient to provide 10 - 12 rounds of irrigation in a cardamom plantation of 0.8 ha. The cost of such a device is estimated to be around Rs.24,000/- (Rs. 16,000/- for excavation work and Rs. 8,000/- for silpauline sheets). Subsidy of 33.33% of the actual cost, limited to Rs.8000/- is provided for the construction of one 200 cu. metre capacity tank to registered small and marginal growers of cardamom. During 2008-09, 65 devices were constructed at total subsidy of Rs.0.026 crores.

The programme for rainwater harvesting implemented in Kerala, Karnataka and Tamil Nadu for small cardamom is replicated for large cardamom in North Eastern states. The terms and conditions and subsidy provided are same as in the case of cardamom (small). During 2008-09, ten rainwater harvesting devices were constructed providing a subsidy of Rs.0.007 crores.

- During the stake holder consultations, the growers expressed the view that irrigation is an area that needs greater attention of the Board in the years to come. **Training programme for quality improvement of spices**

The Spices Board is regularly conducting quality improvement training programmes for all the stakeholders (farmers, officials of state Agri./Horti. Department, traders, members of NGOs) for popularizing the scientific methods of pre/post harvest and storage operations and updated quality requirements for major spices. During 2008-09, training programmes were conducted for 24,852 spices growers in 588 centres and 552 officials of State Agri/Horti. Dept. in 12 centres, 124 traders in 4 centres and 480 representatives of NGOs in 12 centres. Twenty one Regional seminars benefiting 1352 participants were also conducted. The total number of personnel trained under the above programme in 637 centres was 27360.

Organic Farm Certification

The programme aims to help growers/ processors of spices in acquiring organic certification which is a pre-requisite for marketing organic spices.

During XI Plan, the Spices Board has been assisting group of farmers, NGOs and Farmers Co-operative Societies/ Associations in acquiring certification for their farms/ processing units by meeting 50% cost of the certification, subject to a maximum of Rs. 75,000. Individual farmers and processors are eligible for 50% of the cost of certification subject to a maximum of Rs. 25,000 per certification. During 2008-09, assistance to the tune of Rs.0.129 crores has been provided to 22 NGOs / groups and 17 individual farmers covering an area of 3456 ha. benefiting 4556 farmers. Five processing units were also supported under this programme.

During consultation with cardamom growers it was pointed out that they get cheated in the hands of manufacturers of bio fertilizers/insecticides and they stressed the need for a certification system. Also there is excessive use of chemicals (pesticides) that is harmful to both the users and the environment. The growers were of the view that the Board should take steps to publish a list of harmful/banned pesticides as in the case of pharmaceutical industry. The role of awareness generation is also equally important.

Development of Spices in North

Eastern States

Chilli, ginger and turmeric are extensively cultivated in the North Eastern states. Some of the indigenous varieties viz., 'China', 'Nadia', and 'Thingpur' in Ginger, `Lakadong' in Turmeric and 'birds eye' in chilli are considered rich in oil, curcumin content and capsaicin content respectively. The agro–climatic conditions prevailing in NE states are suitable for the cultivation of pepper and large cardamom and these crops can be profitably grown in these regions to create exportable surplus. The Board has identified the potential for promoting production of organic spices in these states by popularizing organic farming practices among the growers to generate exports. The major constraints noticed in the development of spices in NE region are lack of an organised marketing system and lack of knowhow on cultivation and post harvest practices. Therefore, Spices Board implements an integrated scheme for the development of export oriented spices in N.E States with the following components:

Support for Vermi-Compost Units

There is a need to produce organic inputs in the farm itself to maintain soil fertility and to support organic production. In order to enable the growers to produce organic farm inputs, particularly vermi-compost, Rs.2000/- is offered as grant-in-aid to set up a unit with one ton output of vermin-compost. During 2008-09, 2817 vermi-compost units were set up.

Extension Advisory Service

Transfer of technical know-how on production of spices to growers is an important factor in increasing productivity. This programme envisages technical/ extension support to growers on the scientific aspects of cultivation through personal contact, field visits, group meetings and through distribution of literature in vernacular languages for increasing productivity and improving quality of cardamom in Kerala, Karnataka and Tamil Nadu, development of large cardamom in the states of Sikkim and West Bengal and selected spices in the North East and small pockets across the country.

Training of Officers and Farmers of NE States

The Spices Board arranges training programmes for the officers of the state Agri/ Horti. Departments and growers of North Eastern States on the recent advances in the areas of cultivation, harvest and post harvest techniques of spices. The training is arranged in alternate years for officers and every year for farmers. During 2008-09, 88 farmers of the NE states were trained at Indian Institute of Spices Research, Calicut; Kerala Agricultural University, Thrissur; Indian Cardamom Research Institute, Myladumpara and Quality Evaluation Laboratory of Spices Board and spice processing units.

Natural Rubber

Natural rubber, obtained from the latex of the tree, Hevea brasiliensis, is a vital raw material used in the production of around 35000 items of industrial importance. Major portion (50.2%) of Natural Rubber (NR) in India is consumed by the automotive tyre sector, specifically the large tyre sector, the growth of which is directly related to the transportation sector. Hence the demand for natural rubber has been growing with the growth in GDP and industrial growth.

The substitutes of natural rubber are synthetic rubbers (SR) which utilize petroleum products as their feed stock. While the production of SR is highly energy intensive and not an environment friendly process, NR is an environment friendly raw material and renewable resource. For certain specific uses NR is indispensable. For example in products such as aviation tyres and truck tyres in which heat buildup is considerable while in use, NR is preferred because of its property to withstand heat buildup. The synthetic poly isoprene, which can nearly simulate NR, is expensive and in our country it is not produced in the required quantity. It is also not comparable with NR environmentally.

Being a crucial industrial raw material its production in the country has been inward oriented and was mostly meeting the domestic demand. Cultivation of rubber was introduced in the country during the first decade of this century. Rubber is primarily grown in the State of Kerala and adjoining Kanyakumari district of TamilNadu, which are the traditional rubber growing areas of the country. Both areas are geographically and agro-climatically suitable for rubber cultivation. Though rubber was mainly cultivated as large plantations initially, much structural transformation took place in the sector subsequently and now, 89 % of the area and 92% of the production come from small holdings with an average size of 0.50 ha.

However, with focus on globalization the natural rubber production has been increasingly oriented towards major consuming countries in addition to meeting the domestic demand. Also, with the removal of tariff and non-tariff barriers, the natural rubber production has been exposed to heightened international competition. Hence with a view to increasing the production, to meet the growing domestic and export demand, and productivity with a view to be internationally competitive, the Rubber Board has been undertaking a number of promotional measures. As mentioned earlier, rubber is a perennial crop with a gestation period of about 7 years. In terms of yield, there are three stages in the life cycle of rubber plantations; a pre-bearing stage of about 7 years followed by the second stage up to about 14 years where in productivity shows an increasing trend and the last stage after 14th year wherein yield shows a declining trend. Thus viewed, the age structure of the plantations has a crucial bearing on productivity. Hence to maintain higher production and productivity, there is the need for timely replanting of old aged plants and bring new areas under cultivation. Therefore, Rubber Board has been systematically promoting replanting and new planting by providing subsidy and other needed technical supports.

Replanting

Replanting involves substantial initial investment and growers will have to forgo income during the pre-bearing period. Since the small/marginal growers who dominate the rubber sector have difficulty in mobilizing resources to undertake the initial investment and withstand the loss of income during the pre-bearing period, they have been financially supported to carry out the replanting operations with high yielding varieties and to, adopt scientific agro-management practices. In the traditional area of Kerala and Tamil Nadu, a planting grant of Rs. 24500 per hectare, which comes to 20 % of the current cost of cultivation (Rs. 97500/- per hectare), is provided as subsidy for replanting. In the case of nontraditional areas the existing level of subsidy is Rs 22000 per ha which is 25% of the estimated cost of production (paid in 6 annual installments). This financial support is given to only those small growers who own up to 5 hectares of rubber area and the financial assistance is limited to an area of 2 ha only in the traditional region and up to 20 ha in nontraditional areas. Every installment of planting grant will be paid only after ensuring satisfactory maintenance of the holding which is measured using fixed growth parameters and other maintenance indicators.

In the nontraditional regions, considering high cost of planting materials and the cost involved for long distance transportation of all plantation inputs, the following additional incentives are given:

a) Reimbursement of cost of planting materials @ Rs.8.00 per plant limited to cost of 500 plants per ha.ie. Rs. 4000/- per ha (paid in one installment)

b) Transportation grant of Rs. 4000/ per ha for transporting plantation inputs to the site. (paid in 6 annual installments along with the planting grant).

According to the estimates of the Rubber Board, out of the total targeted 48250 ha of replanting during the 11th Plan, only 31750 ha (30000 ha in traditional region and 1750 ha in the nontraditional region) is to be covered under planting grant.

New Planting

Given the need to expand the area under cultivation to meet the growing demand and the limits to further expansion of rubber cultivation in the traditional areas on account of the limited availability of land, new planting is promoted in the non-traditional regions in the states of Karnataka, Orissa, Andhra Pradesh, Goa, Maharashtra etc. The rate of financial assistance and eligibility criteria proposed for new planting is the same as that for replanting.

The 11th plan has a target for new planting 7500 ha in the traditional region and 2500 ha. in the non-traditional areas.

Quality Planting Material

Generation

As a complimentary to replanting and new planting, the Rubber Board has programmes to make available quality planting material. For this purpose, the Rubber Board is maintaining one Central Nursery and 8 Regional Rubber Nurseries in different parts of the country including the nontraditional region other than NE. These nurseries scientifically produce good quality budded plants and are the sources of genuine bud wood material required by private nurseries to produce budded plants. Currently, the Board is in a position to meet only around 10% of the total planting material requirement in the country.

The quality of planting material is a major of concern as it has a bearing on the yield that the grower will actually realize when the tapping starts. There is also hardly any means by which an ordinary small holder can check the quality of planting material at the time of planting.

During the interaction with small holders it was brought out that at present there is hardly any institutional mechanism to check the quality of planting material and they are at substantial risk. This issue also came out during the consultation with planters[2]. If such an uncertainty remains, there appears to be the need for an institutional mechanism that can take the form of certification/licensing of private nurseries (a practice existed earlier) by the Rubber Board.

Supply of Other Inputs

- Rain-guarding – This measure helps to permit tapping of rubber trees during rainy days and as a result of rain-guarding about 15% of increase in productivity can be ensured.
- Plant protection measures – The Board distributes copper oxychloride, copper sulphate and spray oil for spraying mature rubber plantations against the most common disease namely the abnormal leaf fall disease caused by Phytophthora palmivora. Workers have to be trained in the mixing of fungicides and use of sprayers.
- Adoption of scientific tapping practices and application of the stimulant Ethephone, on the bark of rubber trees in limited quantities is proven to increase the yield of rubber.

Soil Protection and Water

Harvesting

There is scientific evidence to prove that yield of rubber can be increased by conserving the rain water falling in plantations there itself by making silt pits. This is also an environmentally sound practice that helps protecting the top soil. But, for the small holders to accept the idea and to adopt this practice, demonstrations in their own fields might be necessary. Demonstrations of soil protection and water harvesting practices in farmers' fields are arranged under RPD scheme. Every year, an area of 4000 ha. (8000 units) in traditional region and 200 ha (400 units) in the NT region are proposed to be covered under this programme during the XI plan. A financial assistance of Rs.3,000/- per hectare is provided for making silt pits @ 250 per hectare in small holdings.

Farmer Group Formation and Empowerment

In a sector where the producers numbering over 10 lakh are small / marginal farmers with very small holding size, a group approach is appropriate to bring in positive changes in the production and productivity scenario. During the last few years the Board has been striving to form farmer groups under the name of Rubber Producers Societies (RPS). The Board is also promoting formation of SHGs in the sector. So far 2184 RPSs and more than 500 SHGs have been formed in the traditional region.

The effectiveness of RPSs in functioning as an extension arm of the Board is well established. Taking into consideration the increasing number of small holdings and the effectiveness of RPS in solving common problems of members, it is felt that more growers should be brought under the umbrella of RPS. Presently only around 25% of the farmers are members of RPS.

The Rubber Producers Societies are being empowered to manage the tiny holdings of its members as a single unit by carrying out cultural operations like planting, fertilizer application, spraying, processing etc. collectively. This will reduce the cost of individual farmers and increase their productivity.

Under the RPS, Self Help Groups of small growers and their family members are also formed. With a view to address the growing labour shortage, RPS / SHGs are encouraged to form labour banks to carry out all types of plantation work and undertake tapping and primary processing of latex, which will be a good vocation especially for women. SHGs can also make use of the ancillary income sources from rubber plantations like honey, rubber and cover crop seed etc. through group activities. SHGs in association with RPS will be able to function effectively in modernizing the small holdings in their area through scientific group management and earn a livelihood for themselves.

The Board supports the RPS and SHG through special projects aimed at the socio-economic development of the resource poor farmers and their families. The objective is to assist farmer groups and SHGs to stand on their own and to function as positive change agents in the sector.

Programmes undertaken under the aegis of RPS with the financial support from the Board include the following.

Farmer education programmes and field training – Farmer education programmes and field training of growers is proposed to be continued as an extension programme to update the knowledge level of the growers in matters relating to planting, harvesting, processing and marketing, with special reference to cost reduction and increase of net farm income.

Group Processing Centres – Community processing centers are proposed to be set up in the RPS sector for processing the latex produced by member growers into good quality sheets. Often the sheets produced by small / marginal growers are poor or inconsistent in quality. Community processing centers provide facilities for processing latex of small growers into superior grades of RSS, thereby fetching better price. Group processing is cost effective and the profit obtained will be shared among the member growers. The GPCs will have facilities for conducting training classes of different types. This facility can be utilized for conducting seminars / meetings / discussions on subjects of common interest to the members of the RPS. Effluent treatment plants are needed for purification of the effluent formed while processing latex into sheet rubber. Biogas will be produced as a result of treatment of the effluent and the gas thus obtained can be used in the smoke house of GPC for drying rubber sheets. The water coming out of the effluent treatment plant is fit for irrigation in neighbouring agricultural fields.

Special Schemes for Non-Traditional Region

The agro-climatic conditions prevailing in the non-traditional region other than NE, comprising the states of Karnataka, Andhra Pradesh, Goa, Orissa, West Bengal, etc. are not ideally suitable for growing rubber. However, certain pockets in this region like South Canara district of Karnataka, East Godavari district of Andhra Pradesh, Mayurbhanj district of Orissa and adjoining areas have been identified as moderately suitable for rubber. It has been proved that under scientific agro-management practices, rubber can be grown as an economically viable crop in these regions.

An innovative scheme implemented in the North East with much socio-economic benefit is the Block Planting Scheme (see Box 3.2). Encouraged by the achievements of block planting scheme implemented by the Board in Tripura, it is being extended to Karnataka and Goa for the exclusive benefit of SC/ST category of growers with the financial support of the concerned State Governments. Other activities like Farmer Group Formation and Empowerment, Soil Protection and Water Harvesting, and Rubber Agro-management Units, in the traditional region are also being taken up in the NT region on a limited scale.

Development Schemes: An Assessment

An overview of the various development schemes for spices and natural rubber implemented by the respective commodity Boards tends to indicate that there is hardly any sphere of production/processing that remains unattended today. Some of the developmental interventions like the formation of Rubber Producers' Societies (RPS) and Self Help Groups (SHGs) are right interventions to help the small producers to tackle various issues that arise on account of the smaller scale of operation. While the Spices Board is known to be associated with Non Governmental Organisations (NGOs) in implementing various schemes, there appears to be scope for emulating the strategy of forming farmers' groups as implemented by Rubber Board. This is likely to be especially useful in those crops where small holders dominate. In cardamom, since there are serious issues regarding the quality of farm inputs like pesticides and bio fertilizers, used in cardamom cultivation the farmers had proposed a certification system for the important inputs.

The replanting and new planting schemes systematically implemented by both commodity Boards have given rich dividends in terms of enhanced production and productivity. It is also much appreciated by the growers. However, there appears to be room for reviewing the subsidy structure not only in terms of the amount of subsidy, but also in terms of the pattern of disbursement. First, it is worth considering the provision of subsidy based on the number of plants grown rather than area planted. Secondly, the subsidy provided shall cover only the interest cost of investment which could be paid directly to the growers' bank account. Thirdly, one could consider a system of "base subsidy" depending on the planting effort(number of plants) and a "supplementary subsidy" on the basis of performance.

Yet another issue relates to the amount of subsidy. As of now subsidy for natural rubber with a gestation lag of seven years is only Rs. 20,000 per hectare (20% of the cost of cultivation) paid in 6 installments. Where as in case of cardamom where the gestation lag is less than three years, the subsidy is as high as 39,171 per ha and Rs.29,675 per ha respectively for the small (up to 4 ha) and marginal growers (4 to 8 ha). paid in two installments. Here it is worth noting that the assistance for planting given in other major rubber producing countries cover a higher proportion of the total cost than in India. The Rupee equivalent of development assistance in Malaysia is Rs.55500/- per hectare, in Thailand it is Rs.54400/- per hectare and in Sri Lanka it is Rs. 55500/- per hectare (Joseph and George 2002). Here it may be noted that assuming as yielding age of 18 years growers pay cess of Rs 54,000/ha. if the yield is 200kg/ha and the cess is at the present level of Rs 1.5/kg. In the current context wherein India has to compete with these countries, the subsidy differential could indeed turn out to be a factor that adversely affects the international competitiveness of Indian producers.

A Wider Issue: On Effectiveness of Replanting Scheme

The basic objective of replanting scheme is to induce the growers to undertake timely replanting such that the share of old aged plants is kept at the minimum. Hence one of the ways to assess the effectiveness of the replanting scheme could be by examining the changes in the age profile of plantations brought about by replanting. Our preliminary enquiry tends.

Research, Development & Extension

R&D and extension is the fuel that drives the engine of agriculture. Given the highly competitive environment in which the plantation sector operates, R&D and extension have assumed importance more than ever before for achieving objectives like cost reduction, productivity enhancement and others. Realizing the importance of R&D and extension, both Rubber and Spices Boards have established R&D Institutes and elaborate extension network to cater to the varied needs of different stakeholders. This background paper examines the R&D and extension activities undertaken at the instance of the two Boards and makes a very preliminary assessment of the present research system using indicators that include among others, R&D intensity and R&D outcomes like the diffusion of the plant varieties developed by these institutes. Instead of giving definite answers focus of this paper is to raise issues that need much detailed exploration and critical analysis for informed policy making.

Spices (Cardamom)

Indian Cardamom Research Institute (ICRI) was established by the Spices Board in 1978 under the Ministry of Commerce and Industry, Government of India, at Myladumpara in the Idukki district of Kerala. ICRI, with other three Regional Research Stations, located at Tadong (Sikkim), Saklespur (Karnataka) and Thadiyankudisai (Tamil Nadu), one satellite Biotechnology laboratory at Cochin and Quality Evaluation Laboratory under it, undertakes research on all aspects of cardamom production. ICRI has 34 scientists of which 29 are with PhD degree and is headed by the Director (Research). A 64 hectares farm is attached to the main Institute at Myladumpara. A farm is attached to each Regional Station for commercial cultivation as well as laying out field experiments.

The primary mandate of ICRI is to develop sustainable production, protection and post-harvest technologies for small and large cardamom (see Box 4. 1 for mandates of ICRI). Its activities have been recently expanded to cover comprehensive research.

ICRI undertakes farmer advisory activities through its Farm Advisory Center manned by a Senior Scientist. These activities include Soil/ Plant test based fertilizer recommendation, pest and disease management advisory services, supply of quality planting material and supply of bioagents.

Soil Health Card is issued to cardamom farmers based on comprehensive soil and plant analysis. Substantial savings upto 25% in fertilizer usage could be achieved in cardamom ecosystem through this service provided by ICRI.

Good Agricultural Practices (GAP) training programme is a residential training programme of three months duration on good agricultural practices for quality spices production to unemployed youths of India who are interested in taking up agriculture as a profession in rural areas. The programme covers various aspects of ecologically sound and sustainable spices production. It may also provide employment opportunities in nursery production, bio-agent production, consultancy services etc. Entrepreneurs / NGOs could utilize the expertise of the trainees for improving their farms. Apart from this, short term training programmes on bioagent production, organic spice production, correct pesticide usage, post harvest handling, nursery practices etc., are also provided.

Weather forecasting and weather based crop advisory services are provided on a weekly basis to cardamom farmers through the Field Offices[1]. In addition to ICRI there are also other actors involved in research on spices. However, it appears that there is much scope for effective interaction among the various institutions/actors. While there is no harm in different institutions getting involved in spices research, there should be adequate opportunities for interaction among the researchers through periodic exchange of research ideas and findings. At the same time, in a situation where resource constraints exist for individual research institutions, it may be worth considering the option of specialized research centres /institutions for major spices or group of minor spices. In any case, what is important is that there should be effective coordination among the institutions/centres involved in similar research.

Achievements of ICRI

ICRI maintains the national conservatory of germplasm of small and large cardamom with the worlds' largest collection of 800 accessions of small cardamom and 300 accessions of large cardamom. This genepool is the base for all present and future crop development programmes in cardamom.

Location specific cardamom clones ICRI-1, ICRI-2, ICRI-5, ICRI-6 and ICRI-7 were developed for Kerala; ICRI-3 for Karnataka; ICRI-4 for Tamilnadu and ICRI Sikkim-1 and ICRI Sikkim-2 for Sikkim and Northeastern Region and released for cultivation at state/ national level. These improved clones have contributed substantially to the increase in the production and productivity of small and large cardamom The fruits of biotechnology were taken to the farm gate level by demonstrating the performance of tissue culture plants of small cardamom in 100 hectares, large cardamom in 100 ha; vanilla in 111 hectares and black pepper in 100 hectares. Many of these beneficiaries were the recipients of productivity awards for the respective crops at national level.

Good Agricultural Practices for mandate crops were evolved in tune with changing ecological, climatic and economic scenario to ensure sustainability of its cultivation.

Cardamom soils have been characterised and agro-ecological zones of Cardamom Hill Reserve (CHR) have been defined for better location specific management of cardamom and other spices.

Integrated pest and disease management strategies were developed which has helped to reduce crop loss to less than 10%. Presently, the research focus is to reduce the toxic chemical load in the crop ecosystem for which effective native isolates of useful microorganisms (bioagents) have been identified. *Trichoderma viridi.*

Commodity Price Instability and Cyclicality

Price Instability: An Issue of Global Concern

The commodity prices in the international market are known for their instability. As pointed out by World Bank (1999), "Commodity prices are notoriously volatile, creating instability and uncertainty for commodity dependent developing countries. It affects governments, producers, traders and local financial institutions, undermines economic growth and skews distribution of income". Price instability and long-term or short–term trends have become major concerns of commodity dependent developing countries.

Price instability creates uncertainty in the markets and this has serious consequences not only in the regional economy but also in the national economy. To be more precise, price instability introduces uncertainty that in turn has negative effect on investment and growth and at the same time adds to the indebtedness agricultural households. Also it adversely affects the income flows of farmers, employment opportunities for the agricultural labourers, which in turn reduction in income of farmers may affect the investment on land and use of fertilizers, adoption of, new and high yielding plant materials, and utilization irrigation facilities, all of which will adversely affect the production and productivity of crops. A study by Deaton and Miller (1995) reported that other things being equal vulnerability to commodity price uncertainty reduces growth rates –the greater the commodity price uncertainty, the slower the country's growth rate and greater the extent of poverty. To address this issue the studies have called for targeted policies at different levels.

Yet another study (Page and Hewitt 2001) notes that commodity price fluctuations, along with the globalization of the world economy and increased liberalization of commodity markets have led to profound changes that seriously affect the weaker economies of the developing world. Commodity price instability has a negative impact on economic growth, countries' financial resources, and income distribution, and may lead to increased poverty instead of poverty alleviation. The world commodity prices being notoriously volatile, driven by changes in global demand and supply, developing countries are particularly affected by external shocks that can result in increased poverty and reduced public funding for health and education.

Addressing Instability

In the literature a number of factors have been identified as contributing to agricultural price volatility in the developing countries. They include factors specific to agriculture like imperfections in information relating to supply and demand side, hedging, and speculation. Perhaps the most important factor in the developing countries relates to supply shocks arising from adverse weather conditions due to limited investment in irrigation and land improvement. Moreover, variations in national income, industrial production and inflation, national liquidity, exchange rate and changes in related policies and cartel behavior have significant role in the volatility prices in the domestic market[1]. Whatever may be the reasons government interventions in the market through measures such as procurement, minimum support price, protection from imports and other initiatives were used in India to minimize the adverse impact of such instability in prices. However, the policy reforms initiated during the last decade in general had the effect of state withdrawal from such active market interventions and leaving the price obtained by the farmers susceptible to market forces more than ever before.

Early attempts to deal with commodity price volatility relied on stabilization schemes set up in the context of international commodity agreements. These arrangements were largely unsuccessful. Globalisation of commodity markets and the lowering of trade barriers, along with priorities focused on sustainable development and poverty alleviation, call for a more innovative approach to commodity risk management. The concerns of all interested parties (traders, governments, producers, farmers) should be taken into account.

Page and Hewitt (2001)[2], among others, show commodity prices have fallen relative to manufactures and services and are likely to continue to do so. There is little scope for increasing the volume of sales sufficiently to counterbalance this. Therefore, the long-term strategy for development for most countries must be to reduce dependence on commodities and move into manufacturing or services. This cannot happen overnight, and the difficulties of adjusting to low and falling commodity prices have been increased by the withdrawal of official support (national and international) for commodities. Financial instruments exist in developed countries for some commodities which allow producers to transfer some of the price risk to the market.

Appropriate assistance to developing countries can help create the conditions for a successful diversification: good physical, social, and institutional infrastructure : transport, communications, health, education, laws, banking system. Developed countries can ensure that their trade policy offers the same treatment to all developing country exports to avoid favouring traditional commodities or discouraging new products and services. There are structural problems of monopoly in commodity markets which require multilateral regulation.

Good institutional infrastructure will also ease the introduction of financial instruments. But using financial instruments has a financial cost, which may be high for poor producers, training and information costs, and institutional requirements. Assistance from countries with financial resources and experience can speed the development of national markets or ease the use of international markets. Any such assistance needs to be linked to a plan for diversification and time-limits to avoid offering a wrong incentive by reducing the expected costs of remaining specialized in commodities.

Some commodities and some types of medium-term fluctuations are not suitable for market-based stabilization. Some countries and some producers within these countries are too poor (or small) to enter the markets, even if they exist (or can be created). The countries which remain dependent on commodities are in general among the poorest. Even those countries which can change their structure will continue to face the problems caused by the trend fall in commodity prices until they have made substantial progress in diversification. Assistance to reduce current poverty must be undertaken alongside assistance to create the economic structure which will reduce poverty more permanently.

Export Orientation

Historically, as in other developing countries, plantation sector in India has been promoted as a foreign exchange earner. In the Indian context, during the import substitution regime characterized by acute shortage of foreign exchange to finance our imports and development, export of plantation crops was especially encouraged. This is one of the major reasons behind the establishment of commodity boards under the Ministry of Commerce for those crops important either from export orientation like cardamom or import substitution like natural rubber. Over the years, a number of changes have occurred in the domestic and foreign market conditions and the economic scenario. Therefore, it is relevant to ask fresh questions about the role of Plantation Sector in the emerging scenario. Some of the important questions are: Has the role of plantation sector in the total export earning changed and if so how? Has there been any inter-crop variation in the extent of export dependence? If there have been a decline in the role these crops, what is the justification for state support schemes? How competitive are these crops in the international market? How effective has been the export promotion measures? These are some of the issues that this paper intends to deal with.

Plantation Exports in India's Merchandise

One of the important dimensions of economic growth is the structural change characterized by a decline in the share of primary agricultural sector and the corresponding increase in the industry and service sectors. Such a structural transformation naturally will also get reflected in the export basket of any growing economy. Indian experience in this respect has not been an exception. With the emergence of a vibrant service sector and growing manufacturing industries, the share of plantation exports in total merchandise trade has declined over the years. As indicated in Table 8.1 in early 1970s the share of plantation sector in merchandise exports was about 13 per cent which was much higher in the pre-1970 period.[1] Needless to say, the export earnings from the plantation sector played an important role in financing

India's imports during our import substitution regime wherein foreign exchange was a major constraint. However, notwithstanding remarkably high growth in the export of plantation commodities, the present share of these crops is only a little over one per cent.

The contribution of the plantation sector in the total export earning has declined it still plays an important role in agricultural exports from the country. As noted by the Rangachari Committee (2002), plantation sector accounted for over 15 per cent of the total agricultural exports earning though they account for only less than one per cent of the total cropped area.

Concentration in Backward Areas

Historically, plantation sector has been considered as an instrument of modernization, as it helped in opening up of the underdeveloped areas and creation of social overhead capital. However, even today there are evidence to indicate that limited access to needed infrastructure in the plantation areas leads to increased cost of production and stands in the way of enhancing international competitiveness. Viewed in the similar vein the infrastructure deficit also undermines the ability of plantation sector to withstand import competition in the context of removing various barriers to trade.

From the point of view of regional economies, they are a major source of livelihood for their populations in certain less developed regions. Besides, crops like tea, coffee, cardamom and pepper are cultivated in ecologically fragile areas and the sustenance of plantation agriculture is important for the long-term development of such regions. The survival strategies of the planters under the new era of heightened international competition and looming crisis is shown to have had adverse effect on the fragile eco-system of these areas (Damodaran, 2002)

Export Intensity: Inter-crop Variation

While there has been a declining share of plantation crops in the country's merchandise exports, there is significant variation in the observed level of export intensity across different crops. As far as spices are concerned, depending on the trend in export intensity, they could be divided into two categories. The first category comprises of those crops wherein export intensity remained stagnant or declined and the second group includes those commodities that recorded an increase in export intensity[2]. Among the 10 major spices for which data is available, export intensity declined or remained stagnant in case of five crops (black pepper, small cardamom, ginger, garlic and fennel) during the post 1990 period (see Table 8.4). Though export intensity declined for some of the crops like black pepper, their export intensity is remarkably high at over 50 per cent at present and in some years it was as high as 70 per cent. In the case of small cardamom, while export accounted for over 70 per cent of the domestic production in the 1970s, present export intensity is only around five per cent. This drastic decline in the export intensity despite more than three fold increase in domestic production needs to be seen in the context of growing domestic market and effective market promotion by the Spices Board. Of course, the emergence of low cost producing countries like Guatemala also had its influence.

Export-Oriented Promotional Activities of the Spices Board

With a view to enhance the acceptability of Indian spices in the world market, apart from the export promotion activities like participation in trade fairs, advertisement through the distribution brochures, setting up of spices parks and quality certification and quality improvement measures, a host of programs at the production level are also implemented. Some of these programmes are being implemented in association with the NGO. These programmes include, but not limited to the following:

Improved Cardamom Curing Devices

Cardamom is dried in traditional curing houses using firewood as fuel. Sun drying is not popular due to the loss of green colour during the process. As the productivity and production has been registering an upward trend, firewood requirement for curing cardamom has been increasing. As the wind fallen trees are not enough to meet the growing demand of firewood, the growers are forced to meet their requirement of firewood, either from the market or from resorting to cutting trees leading to degradation of forest cover. Moreover, wood is required for the construction of new curing houses especially for racks to spread cardamom and to provide false ceiling in the curing houses to preserve the heat.

Few innovative growers have started installing cardamom curing systems using alternate fuels, viz. Diesel, LP Gas which give their produce better colour and cost effective curing. These driers are eco-friendly, labour saving and easy to operate. The drying time is reduced from 28 – 36 hours to about 20 hours in these driers. The objective of the programme is to popularize the cardamom drying systems using LPG/ Diesel/Biomass and firewood as alternate fuel among the small growers of Kerala, Karnataka and Tamil Nadu by providing 33.3 % of the actual cost of drier as subsidy subject to a maximum of Rs. 60,000/- for one device. The non-subsidy portion is met by the growers from their own funds or through institutional finance. The Spices Board has prepared a list of approved suppliers and has fixed upper ceiling cost for driers of various capacities. The driers should be purchased from approved suppliers. During 2008-09, the Board had assisted for installation of 83 improved cardamom curing devices at a financial outlay of Rs.0.375 crores.

During consultation with the farmers, it was reported that some improvement to the traditional curing house has been introduced by the growers that in turn results in about 50 per cent savings in fire wood. The growers demanded that the Board should have provisions to encourage such innovations at the farm level.

The large cardamom growers traditionally cure their cardamom in the locally fabricated bhatties. This does not ensure proper drying and ideal colour to the cured cardamom. The Board had introduced and evaluated a number of curing methods using different fuels and has selected a system which gives the best quality. In order to popularize this method, the Board is providing subsidy @ Rs.5,000/- for 200 kg capacity and Rs.9,000/- for 400 kg capacity drier respectively.

Promotion of Organic Cultivation

Internationally, the niche market for organically produced spices is growing at a fast rate. Early entry into this segment will improve the exportability and demand for Indian spices. Further, availability of organically grown spices will help the country to withstand competition from low cost countries in south East Asia. The major bottlenecks in promoting organic farming are non-availability of organic farm inputs and high cost of organic certification of farms and processing units. In order to promote organic production of spices, programmes like organic farm certification assistance, support for setting up of vermi-compost units, and promoting organic cultivation of spices are being implemented by the Board.

Growers were of the view that the contents of the package of practice are increasingly influenced by the traders which in turn leads to indiscriminate use of fertilizers as well as pesticides. This is harmful not only to the workers and environment but also to the consumers. While the growers are increasingly moving towards bio fertilizers, there is increasing sale of spurious fertilizers. Growers felt that if the Spices Board could specify the harmful pesticides and introduce a system of fertilizer certification, exploitation of growers in the hands of traders/manufacturers could be reduced.

Construction of Drying Yards for Spices

This programme aims at construction of concrete/cement drying yards in the farmers' field for drying spices under hygienic condition. The estimated cost of construction is Rs.600/ - per sq.mtr. A subsidy of Rs.150/- per sq.mtr is provided towards 25% of the cost of construction for general category of growers. Scheduled caste and tribes are offered higher subsidy of Rs.300/- and Rs.540/- per sq.m towards 50% and 90% of the cost of construction respectively.

In addition, construction of community drying yards by NGOs, farmers groups (SHG) and association of farmers and Agriculture Market Committees are supported by providing subsidy @ 50% of cost of construction. Individual growers are provided subsidy up to 150 Sq.Mt. of drying yard and NGOs, Farmers Associations, Agricultural Market Communities are eligible for assistance proportionate to the members thereof. During 2008-09, drying yards of 23,254 sq.m were constructed for pepper, ginger, turmeric, chilli and seed spices at a total expenditure of Rs.0.249 crores.

Spices Parks

Spice Parks provide all the facilities needed for a vibrant sector oriented towards both export and domestic market under one roof thus effectively reap economies of scale and scope. The facilities provided in the parks include, but not limited to, facilities for primary marketing (e -auction in case of cardamom), processing, certification, finance and point of final sale.

Sum up

With the emergence of a vibrant service sector and growing manufacturing industries, the share of plantation exports in total merchandise trade declined over the years. In 1970s the share of plantation. sector in merchandise exports was about 13 per cent; currently the share of these crops appears rather negligible with a little over one per cent of total merchandise exports, though their share in total agricultural exports is as high as 15 per cent with less than one per cent of the total cropped area. However, the importance to the national economy at present is more than ever before on account of its contribution towards achieving inclusive growth as this sector is highly labour (women) intensive, operates in the backward and ecologically fragile regions and is dominated by the small holders whose livelihood is linked to this sector (Joseph 2010c).

Notwithstanding varied measures initiated in the past, the export orientation of some of the crops like small cardamom and black pepper have been declining or remaining constant. It is further evident that with the exception of a few crops, export intensity is at a lower level, which points towards the further scope for increasing export earnings from these crops. Though the Spices Board has initiated a number schemes like organic cultivation, support facilities for hygienic production, ensuring quality standards and setting up of spices sparks, a large number of crops are remaining out of the ambit of export oriented production/ processing schemes. Hence there is a strong case for extending the export oriented production measures to other crops that are today remaining as "orphan crops". But, as already noted, an efficient system of domestic supply and production conditions is a pre requisite for effective export promotion. However, in case of spices, there is a disjuncture between production and export and hence the current strategy of "harvesting without sowing" could turn out to be unsustainable.

While the spices like cardamom are known for their export orientation, natural rubber, being an industrial raw material with fast growing domestic demand, has been an import substituting crop. Though, export of natural rubber is recent phenomenon, its export intensity at present is comparable to that of cardamom. Given the open economy context and excess demand situation for NR in the domestic market, perhaps the focus of policy measures and institutional interventions needs to be focused to further enhancing the domestic productivity to withstand the potential import competition from other countries. This will also help the sector to meaningfully enter into the world market.

Export Competitiveness and Asean-India FTA

Given the current context wherein, man-made restrictions on trade are done away with under globalization, and a large number of RTAs and PTAs are being signed by India, at the national interest, has become a point of concern for the plantation sector . In India, the plantation sector appears to assume more importance today in the national and regional economies (like Kerala, Tamil Nadu, Karanataka and the North Eastern States) than ever before. While the importance on account of its significant contribution to foreign exchange has been highlighted in the yesteryears, its role in achieving inclusive and environmentally sustainable growth is gaining greater attention today.

Historically plantation crops have been raised in large estates. Over the years there has been a transformation in the ownership structure and that today there are over 4 million growers mostly small and medium. The sector employs over 1.7 million plantation workers and nearly 2 crores farm workers mostly women. Though this sector accounts for only about one per cent of the total cropped area it accounts for over 15 percent of the agricultural exports. At the same time the plantation sector is located mostly in the regions that are more backward than the least developed countries that were exempted from tariff reduction under WTO. The prices of plantation commodities are known for their volatility and cyclicality and that got accentuated under WTO. If the available empirical evidence is any indication, as the policy pendulum shifted from protection to greater competition under the aegis of WTO there has been simultaneous decline in the price of most of the crops. (Joseph and Joseph 2005) along with heightened volatility in prices (Joseph 2004, Subrahmanian 2008) that induced farmers to resort to such survival strategies that are hostile to ecology (Damodaran 2002) and inhibited investment in plantations (Government of India 2006). These and other related developments made the plantation sector, which generates massive rural employment opportunities and located largely in economically backward and ecologically fragile regions, unviable. In the context of accumulated debt burden beyond the paying capacity many farmers engaged in plantation sector opted to take their lives leading to farmers' suicide in India at a rate unheard in human history (Mohanakumar and Sharma 2006, Jeromi 2007, Nair and Menon 2009, Reddy and Mishra 2009 and others)

The likely impact of the FTA on plantations could be analyzed either using a simple static analytics or a dynamic perspective tempered by reality. Those opposing FTA, by adopting the former approach, fear increased imports on the following grounds. With the ongoing Initiative for ASEAN Integration (IAI) and other measures to level up the development divide among ASEAN, Cambodia, Laos and Myanmar are likely to record a better economic performance in the years to come as being currently experienced in Vietnam. In such a context, facilitated by FTA, India's export of manufactures and capital goods to these countries would increase. This could further worsen their trade deficit as in case presently with Vietnam where India has a trade surplus of over one billion. In such an eventuality these countries would be left with hardly any option but to increase their exports to India at any cost. Being countries with primary sector focus and agro-climatic conditions similar to Southern states like Kerala their primary exports would be competing with crops from these states. Competition from Vietnam has already been felt with its full intensity.

In the similar vein there is a perception of heightened import competition from other ASEAN countries as well. Compared to India, old ASEAN countries like Malaysia, Indonesia and Thailand are having much higher production of plantation commodities like black pepper, natural rubber, coffee and to a lesser extent tea. Thanks to large and growing domestic market, production of these crops in India is increasingly for the domestic consumption. Production of plantation crops in ASEAN, on the other hand, is mostly for the world market as their domestic market is very small. Yet, India accounts for only a negligible share of their total exports while these exports form a substantial part of India's imports (see table 9.1). To illustrate, India accounts for only about 3.8 per cent of Vietnam's export of pepper while their share in India's import is as high as 27 percent. Similarly, India provides only 1.9 per cent of the export market for Thai natural rubber but it accounts for over 45 per cent of India's imports. Hence given the large domestic market of India and large scale supply of these commodities in the ASEAN countries the removal of tariff barriers, it is feared, would result in increased import competition and leading to fall in domestic prices. The solution lies in doing away with FTA such that the plantation sector could be fortified from external competition. Such simple static analytics might appear appealing for those opposing FTA .

From Perception to Reality

To appreciate the reality we need to move out of the simple voice of dissent against FTA that envisages no tariff reduction in most commodities and only a phased reduction for a few others. If the real concern is with plantations and the millions involved therein, the attempt should be to correct the anomalies in the import regime and help developing a vibrant processing sector that provides growth impulses to the country's plantation sector. Given this sector's role in inclusive growth and operates in ecologically fragile regions that are more backward than some of the least developed countries that were exempted from tariff reduction commitment under WTO, we don't need a Kautilya to make the case for subsidizing this sector. Let us now turn to the issue of imports and domestic prices. We have seen that after India-Thailand FTA there has been a massive increase in imports of natural rubber. But this import was not associated with any decline in prices. Instead the price of natural rubber recorded nearly a three-fold increase – a rate of increase unheard in the history of rubber. Similar instances could be multiplied. The observed pattern needs to be seen against the reality wherein domestic prices are governed by prices in the world market which in turn is shaped by global demand-supply conditions and other factors.[2] In such a context, it is rather myopic to argue that domestic prices are determined by imports. It is not to argue that imports have no bearing on domestic prices. In fact even a threat of imports is sufficient to cause short run price fluctuations.

To be fare, threat of such import competition has already been taken into account in the FTA. Hence some of the products are permitted to enjoy the ongoing level of protection (negative list) while in others tariff reduction has been effected in a phased manner (special products) such that productivity and competitiveness could be built up. More importantly, since the share in India's total import is more than three per cent in all the partner countries, India could resort to safeguard measures in case of substantial injury for the local producers. What is needed is an effective mechanism to monitor prices and markets for which there is no dearth of institutional mechanism at present.

For those opposing FTA since tariff cut under FTA would lead to reduction in protection, tariffs needs to be maintained to shield the domestic industry from import competition. It is productivity along with cost of production and exchange rate that determines the required level of tariff. Table 9.2 presents data on productivity (production per hectare) in competing

ASEAN countries as percentage of India's productivity. On the basis of observed productivity difference between India and competing countries in some of the crops, it is evident that even if India reverts to the tariff levels in pre-liberalization period these crops could not be protected to the level of their satisfaction.

Importance of India for ASEAN exports and their share in India's Import for select Plantation Commodities

Crops	2000		2007	
	Share of India in total export	Share in India's import	Share of India in total export	Share in India's import
Vietnam				
Coffee	0.22	28.34	0.24	13.65
Tea	1.31	9.64	0.52	2.39
Pepper	1.05	11.31	3.79	26.98
Natural rubber	0.41	10.12	0.51	2.79
Malaysia				
Coffee	0.07	0.04	0.92	0.09
Tea				0.00
Pepper	1.01	7.33	0.24	0.40
Natural rubber	0.30	29.70	0.52	4.66
Thailand				
Coffee	0.01	0.08	1.83	0.99
Tea			0.01	0.00
Pepper			0.01	0.00
Natural rubber	0.13	28.94	1.93	45.59
Indonesia				
Coffee	19.24	54.64	15.65	36.69
Tea		47.30	32.66	11.89
Pepper	67.62	31.13	72.36	43.58
Natural rubber	3.43	9.20	17.14	38.95

Source: The World Bank and UNCTAD, COMTRADE database provided as part of World Integrated Trade Solution (WITS)

Concluding Remark

ASEAN India FTA could be celebrated as a landmark in India's integration with Asia and a beginning of an end to the isolation from major trading blocks. FTA could also be considered as part of India's strategy, in the context of newly emerging markets, to diversify her trading partners with greater role for developing countries and an indication of her commitment to south-south cooperation. The FTA, however, had to face vociferous voice of dissent from some sectors in India. In this paper we have undertaken a preliminary analysis of the likely impact of the pact with focus on plantation sector.

The extent of gain and loss to the parties involved vary in most agreements. ASEAN India FTA is no exception. As Jairam Ramesh (2008) rightly observed, along with macro gains there will be micro pains to be felt by specific industries, specific livelihoods and specific regions. Here, it is not denied that given the present state of plantations, it has the potential to be one such affected sector. However, we have argued that the additional threat posed by FTA is not substantial when compared to the threat that the sector has been confronting over the years under the trade policy regime that has the provision for duty free import. Given the current import policy, import is not necessarily caused by tariff reduction and contrary to the common belief, import do not determine domestic price which, in reality, is shaped internationally. Since integration leaves scope for both cooperation and competition, there is the potential to collectively address the commodity problematic by harnessing the new possibilities for coordinated action to influence the commodity prices. But for any such actions to be effective and sustainable there is the need to bridge the efficiency gap in the sector across countries. Thus viewed, with or without FTA, the plantation sector cannot afford to continue in the current form as economic history reveals that no country has managed to cross $500 per capital income by resorting to the primary production as it exists in India's plantation sector. While the trade liberalization envisaged in plantation sector under the FTA is gradual and in a phased manner, there has to be a concerted effort towards evolving a vibrant system of innovation and production. This would call for addressing various issues in the sphere of production, processing and marketing, labour and employment, agricultural research and extension, diffusion of innovations including the application of new technologies and not to speak of the development of much needed general and specific infrastructure, and human capital for the development of plantation sector and above all a coordination between different dimensions.

Guidelines for the Central Sector Scheme Monitoring of Pesticide Residue at National Level

The Department of Agriculture and Cooperation, Ministry of Agriculture is regularly monitoring the presence and levels of pesticide residues in food commodities and environmental samples under the central sector scheme, "Monitoring of Pesticide Residues at National Level". The scheme was initiated during 2005-06 and is being implemented through Indian Agricultural Research Institute, New Delhi and has **23 participating laboratories (Annexure-I)** from the Ministry of Agriculture, Indian Council of Agriculture Research, Ministry of Health and Family Welfare, Ministry of Environment and Forest, Council of Scientific and Industrial Research, Ministry of Chemical and Fertilizers, Ministry of Commerce and State Agricultural Universities.

Objectives and Targets

- To test residues of pesticides, their metabolites and other related contaminants in food commodities and environmental samples like soil and water across the country.
- To identify crops and regions in the country having preponderance of pesticide residue contamination in order to focus extension efforts for Integrated Pest Management (IPM) and Good Agriculture Practices (GAP) aimed at judicious use of pesticides.
- To strengthen infrastructure at Plant Quarantine Stations to assess the contamination of pesticide residues in imported food commodities.
- To check and certify presence of pesticide residues in food for export.

The projected targets for the scheme for XII Plan (2012-2017) are at

Annexure-II

Pesticides residue data generated under the scheme are shared with State Governments and concerned Ministries/Organizations to initiate corrective actions and awareness building among farmers.

The Scheme is monitored by two Committees viz. Steering Committee and Technical Committee. The policy decisions related to the scheme are under the purview of the Steering Committee headed by the Joint Secretary (Plant Protection), Department of Agriculture & Cooperation (DAC) while the Technical Committee headed by the Assistant Director General (Plant Protection), Indian Council of Agriculture Research (ICAR) takes care of the technical issues related to the scheme including purchase of equipments, assessment/identification of laboratories and technical work plan. The Project Coordinating Cell of All India Network

Project (AINP) on pesticide residues is the nodal centre for implementation of the scheme. The Network Coordinator, AINP, IARI, New Delhi is the Member -Secretary of the Committee and is the nodal person for the day to day implementation of the scheme related to financial and technical matters and submission of periodical reports, annual report and need based information to the DAC. Participating laboratories are provided with highly sophisticated analytical equipments like, Gas Chromatograph(GC), High Performance Liquid Chromatograph (HPLC), Gas Chromatograph-Mass Spectrometer(GC-MS) and Liquid Chromatograph-Mass Spectrometer (LC-MS).

Participating laboratories collect samples of various food commodities such as vegetables, fruits, cereals, spices, pulses, milk, butter, irrigated water, fish, meat, tea etc. from across the supply chain such as Agriculture Produce Marketing Committee (APMC) markets, retail points and Public Distribution System outlets (PDS). Irrigated water and soil samples are also collected from agricultural fields across the country. The samples are then analyzed for the presence of pesticides residues.

During the period (October, 2006 to March, 2013) out of the 84,486 samples of various food commodities such as vegetables, fruits, cereals, spices, pulses, milk, butter, fish, meat, tea, honey etc. and environmental samples like soil and water that were collected and analyzed, 1487 (1.76%) samples were found to contain pesticide residues above MRL.

During 2012-13, a total of 16,494 samples of food commodities including water have been analyzed, of which 436 (2.6%) samples were found to contain residues above MRL. Status of Pesticide Residues in Various Food Commodities and environmental samples is at **Annexure-III.**

The laboratories involved in the scheme have been chosen from the existing network of laboratories under various ministries, departments and institutes on the basis of their technical expertise in the area of pesticide residues analysis and their geographical location in order to represent different agricultural zones in the country. The commodities to be monitored by each laboratory are location specific with special emphasis on those widely consumed there. It is proposed to associate private laboratories accredited by the National Accreditation Board for Testing and Calibration Laboratories (NABL) and certain other laboratories functioning under Government Departments in the scheme to increase the coverage of monitoring of pesticides residues crop-wise, region wise and pesticide-wise.

Total financial outlay required to meet the expenditure for this scheme during XII Plan period (2012-2017) is Rs. 5009.37 lakhs and year-wise budget allocation is as under. (Rs. in lakhs)

	2012-13	2013-14	2014-15	2015-16	2016-17	Total
Non-recurring	100.00	300.00	300.00	400.00	400.00	1500.00
Recurring	676.474	694.474	699.474	734.474	704.474	3509.37
Total	**776.474**	**994.474**	**999.474**	**1134.474**	**1104.474**	**5009.37**

It is pertinent to mention that no land is needed. No construction of building is envisaged in this scheme. The cost estimate provided include 10-15% escalation in cost owing to variable factors viz. purchase of equipments , travelling costs , wage bills, cost of consumable like chemicals, glassware, training component, miscellaneous and incidental expenses of intermittent nature. However, new equipments like Gas chromatography-mass spectrometry (GC-MS)/ Liquid chromatography-mass spectrometry (LC-MS)/ Gas chromatography (GC)/ High performance liquid chromatography (HPLC), Uninterruptible power supply (UPS)/ Generator, homogenizer, rotary evaporator, air compressor is proposed to be procured at a cost of Rs. 1500 lakhs. A sum of Rs 2032.62 lakh is proposed for engagement of contractual manpower in the scheme. The year-wise outlay is given in **Annexures –IV, IV (a) & IV(b)**

There is no proposal to create new posts. However, contractual manpower for residue analysis will be hired on the ICAR pattern after the approval of Steering Committee.

A sum total of Rs 2032.62 lakhs is proposed for engagement of contractual manpower in the scheme. Rs. 337.62 lakhs is proposed as cost towards salary for contractual services engaged in NPQS, Rangpuri and RPQSs, Mumbai & Chennai and Rs. 1695.0 is proposed under Grants-in-aid towards salary for contractual services.

CHAPTER-VII

Summary and Observation

The cardamom cultivation have been adopted in organic method which is partially adopted by majority of Growers. There is a proportionate range and ratio of demand and supply for Indian cardamom. In the context of institutional innovations as manifested by the rules and laws governing marketing along with technological innovation as indicated by the introduction of e-auctions, this book explored the varied spaces of exclusion that exist in the current system of marketing. With a view to locating the spaces of exclusion the study has made use the taxonomy of social exclusion by Amartya like passive and active exclusion vis-a-vis the instrumental and constitutional exclusion. This revealed a conceived new plausible categories exclusion like subordinated or unequal inclusion, illusive inclusion and sustained exclusion. It was observed that in the conventional auction system, active exclusion has been prevalent on account of the practice of taking a fixed quantity as sample from all the lots regardless of its size and delay in payment along with the absence of title deeds for large number of growers.

Conclusion

Indian Cardamom has a full of uniqueness for which Quality and desired demand by world consumers. At the same time small growers were severely affected in the way of price fluctuating, unable to maintained crop in order to price raise for inputs and manures. Therefore, this is inevitable for shifted to organic cultivation of cardamom in this scenario.Auction system may be favors to small farmers according to their quality of Goods. Either Government or Spices Board must concentrate to cultivation point of view and export point of view in order to safeguard of small growers and fulfillment of foreign exchange through high export.

ANNEXURES

Annexure-I

List of the participating laboratories under the central sector scheme, "monitoring of pesticide residues at national level"

1. Project Coordinating Cell, All India Network Project on Pesticide Residues, LBS Building, Indian Agricultural Research Institute, New Delhi
2. Dept. of Entomology, Punjab Agricultural University, Ludhiana, Punjab
3. ICAR Unit No.-9, BTRS Building,Anand Agricultural University,Anand
4. Dept. of Entomology, Mahatma Phule Krishi Vidyapeeth, Rahuri, Maharashtra
5. Dept. of Entomology, College of Agriculture, Kerala Agricultural University, Vellayani, Kerala
6. Division of Soil Sci. &Agril. Chemistry, Indian Institute of Horticulture Research, Hessaraghatta Lake Post, Bangalore,Karnataka
7. Dept. of Entomology, Rajasthan Agricultural University, Research Station, Durgapura, Jaipur
8. Acharya N.G. Ranga Agricultural University, E.E.I. Premises, Rajendranagar, Hyderabad, Andhra Pradesh
9. Dept. of Agricultural Entomology, Tamil Nadu Agricultural University, Coimbatore ,Tamil Nadu
10. Institute of Pesticide Formulation Technology, Sector–20, UdyogVihar, Gurgaon, Haryana
11. National Institute of Occupational Health, P. B. No. 2031, Meghani Nagar, Ahmedabad, Gujarat
12. Western Region Referral Laboratory, Department of Veterinary Public Health, Bombay Veterinary College, Parel, Mumbai, Maharashtra
13. MPEDA, MPEDA House, Panampilly Avenue, Kochi , Kerala
14. Pesticide Toxicology Laboratory, Indian Institute of Toxicology Research, Mahatma Gandhi Marg, Lucknow, Utter Pradesh
15. Trace Organic Laboratory, Central Pollution Control Board, PariveshBhawan, East Arjun Nagar, Delhi
16. National Environmental Engineering Research Institute, Nehru Marg, Nagpur, Maharashtra
17. Regional Plant Quarantine Station, Haji Bunder Road, Sewri, Mumbai, Maharashtra

18. Regional Plant Quarantine Station, G.S.T. Road, Meenambakkam, Chennai , Tamil Nadu
19. AINP on Pesticide Residues, Directorate of Research, Research Complex Building, Kalyani, Nadia, West Bengal
20. Dept. of Entomology, Dr. Y.S.P. Univ. of Horticulture & Forestry, Nauni, Solan, Himachal Pradesh
21. National Plant Quarantine Station, New Delhi
22. National Institute of Plant Health Management (NIPHM), Pesticide Management Division, Rajendranagar, Hyderabad
23. Central Agriculture Research Institute (CARI), Port Blair, Andaman and Nicobar

Annexure-II

TECHNICAL PROGRAMME (2012-2017)

Projected samples to be analyzed

S.No.	Commodity	2012-13	2013-14	2014-15	2015-16	2016-17	Total
1.	Vegetables	6000	7200	7800	8100	8100	37200
2.	Fruits	2000	2800	3000	3100	3100	14000
3.	Rice	1000	1100	1200	1400	1400	6100
4.	Water	4000	4200	4400	4500	4500	21600
5.	Fish/crustacean/ meat/eggs	1600	1725	1725	1800	1800	8650
6.	Wheat	800	825	800	900	1000	4325
7.	Milk	650	650	650	700	700	3350
8.	Butter	650	650	650	650	650	3250
9.	Spices	550	550	550	550	550	2750
10.	Tea	300	300	300	300	300	1500
	Total	17550	20,000	21075	22000	22100	102725
Private NABL accredited laboratory							
Pesticides/aflatoxin/heavy metal/antibiotics in food commodities & environmental samples		1900	2000	2100	2200	2200	10400
						Total	**113125**

Annexure-III

Status of Pesticide Residues in Various Food Commodities and environmental samples:

During the period April 2010 to March 2013, commodities which have frequently showed the presence of pesticides residues above MRL are given below:

Commodity	Samples analyzed	Samples above MRL	Pesticides which are found Above MRL	Commonly detected Non approved Pesticides
Vegetables	18,704	540 (2.9%)	Chlorpyriphos, Cypermethrin, Phorate and Ethion	Acephate, Profenophos, Quinalphos, Imidacloprid and Triazophos
Fruits	6,263	54 (0.86%)	Chlorpyriphos	Quinalphos, Profenophos and Cypermethrin
Spices	1,433	158 (11%)	Quinalphos and Cypermethrin	Dithiocarbamates, Profenophos, Chlorpyriphos and acetamiprid
Tea	566	7 (1.2%)	Ethion, Dicofol	Cypermethrin
Rice	2,422	61 (2.5%)	Chlorpyriphos and Quinalphos	Acephate and Cypermethrin
Wheat	2,107	42 (2%)	Cypermethrin	Parathion methyl and Quinalphos
Pulses	2,171	3 (0.14%)	Cypermethrin	Chlorpyriphos and Malathion

Annexure IV

XII Plan Year-wise Budget Outlay for the Scheme

"Monitoring of Pesticide Residues at National Level"

Consolidated (2012-2017) **(Rs. in lakh)**

Code & sub-code	2012-13	2013-14	2014-15	2015-16	2016-17	Total
Computer Code-2401- Crop Husbandry						
107-Plant Protection 12-MONITORING OF PESTICIDE RESIDUES AT NATIONAL LEVEL						
01.12.12 – Foreign Travel Expenses	3.5	3.5	3.5	3.5	3.5	17.50
13 – Office Expenses	17.00	20.00	25.00	25.00	25.00	112.00
16 - Publication	0.20	0.20	0.20	0.20	0.20	1.00
20 – Other Administrative Expenses	1.25	1.25	1.25	1.25	1.25	6.25
21 – Supply & Material	20.00	20.00	20.00	25.00	25.00	110.00
28 – Professional Services	0	0	0	30.00	0	30.00
30 – Contractual Services	67.524	67.524	67.524	67.524	67.524	337.62
31 – Grants-in-Aid	649.00	864.00	864.00	964.00	964.00	4305.00
52 – Machinery & Equipment	15.00	15.00	15.00	15.00	15.00	75.00
Travel Expenses (T.E.)	3.00	3.00	3.00	3.00	3.00	15.00
Total	**776.474**	**994.474**	**999.474**	**1134.474**	**1104.474**	**5009.37**

Annexure-IV (a)

BUDGET OUTLAY YEARWISE – XII Plan: Grants-In-Aid

(Total amount in lakh)

S.No.	Head	2012-13	2013-14	2014-15	2015-16	2016-17	Total
A.	Non Recurring	100	300	300	400	400	1500
B. Recurring							
	Pay (Contractual Services)	339	339	339	339	339	1695
	Travel Expenses	10.0	25.0	25.0	25.0	25.0	110.0
	Contingency	200.0	200.0	200.0	200.0	200.0	1000.0
	Sub-Total	549	564	564	564	564	2805
Total(A+B)		649	864	864	964	964	4305

The amount of Grants-In-Aid is to be given to the IARI for making it available to the centers other than that of the Directorate of PPQ&S, NH-IV, Faridabad.

Annexure IV(b)

(Total amount in Lakh)

S. No	Name of PQ Station	Name of Post	No. of Post	Emolument	2012-13	2013-14	2014-15	2015-16	2016-17	Total
1.	NPQS, Rangpuri, N. Delhi	Reasearch Associates(As per ICAR norms based on Ph.D. or M.Sc.)	4	24000 + 30% HRA = **31,200/**	14.976	14.976	14.976	14.976	14.976	**74.88**
			1	23000 + 30% HRA = **29900/-**	3.588	3.588	3.588	3.588	3.588	**17.94**
		Accountant	1	12000	1.44	1.44	1.44	1.44	1.44	**7.20**
		Computer Operator	1	12000	1.44	1.44	1.44	1.44	1.44	**7.20**
		Lab attendant	2	6000	1.44	1.44	1.44	1.44	1.44	**7.20**
	RPQS,	Reasearch	1	24000 +	3.744	3.744	3.744	3.744	3.744	**18.72**

2	Mumbai	Associates (As per ICAR norms based on Ph.D. or M.Sc.)		30% HRA **=31200/**						
			3	23000 + 30% HRA **=29900/**	10.764	10.764	10.764	10.764	10.764	**53.82**
		Senior Research Fellow (SRF)	1	12000 + 30% HRA **= 15600/-**	1.872	1.872	1.872	1.872	1.872	**9.36**
		Accountant	1	15600 (consolidated)	1.872	1.872	1.872	1.872	1.872	**9.36**
		Store keeper	1	15600 (consolidated)	1.872	1.872	1.872	1.872	1.872	**9.36**
		Lab attendant	2	7500 each (consolidated)	1.80	1.80	1.80	1.80	1.80	**9.00**
3.	RPQS, Chennai	Reasearch Associates (As per ICAR norms based on Ph.D. or M.Sc.)	3	24000 + 30% HRA = **31,200/-**	11.232	11.232	11.232	11.232	11.232	**56.16**
			1	23000 + 30% HRA **=29900/**	3.588	3.588	3.588	3.588	3.588	**17.94**
		Senior Research Fellow (SRF)	1	16000 + 30% HRA **= 20800/-**	2.496	2.496	2.496	2.496	2.496	**12.48**
		Accounts Operator	2	12000 each (consolidated)	2.88	2.88	2.88	2.88	2.88	**14.40**
		Lab attendant	3	7000/- each (consolidated)	2.52	2.52	2.52	2.52	2.52	**12.60**
		TOTAL			**67.524**	**67.524**	**67.524**	**67.524**	**67.524**	**337.62**

References

1. Joseph, K.J. (2010), 'Inclusive Innovation Systems in Plantation Sector: The State of Database and its Limits,' NRPPD Discussion Paper,Centre for Development Studies.
2. The World Bank, (2006), 'Enhancing Agricultural Innovation: How to Go Beyond the Strengthening of Research Systems,' http://siteresources.worldbank.org/INTARD/Resources/Enhancing_Ag_Innovation.pdf
3. Unpublished project reports on "Scouting and documentation of grassroots innovations" of PDS (2010 and 2011).
4. Bhalla, G.S. (2004), "Globalisation in Indian Agriculture", Vol. 19, State of the Indian Farmer: A Millennium Study (Ministry of Agriculture, Government of India and Academic Foundation, New Delhi).
5. Nair, K N and Vineetha Menon. (2009), "*Distress, Debt, and Suicides among Farmer Households: Findings from Village Studies in Kerala.*", In Agrarian Crisis in India, ed. D. Narashimha Reddy, and Srijit Mishra, New Delhi: Oxford University Press.
6. Rajasekharan, P. and Veeraputhran, S. (2002), Adoption of intercropping in rubber smallholdings in Kerala, India: A tobit analysis. Agroforestry Systems, 56 (1): 1-11.
7. Raju Sunitha, (2010), "*ASEAN- India FTA: Emerging Issues for Trade in Agriculture*", International Seminar on ASEAN India FTA and Way Forward, NRPPD, Centre for Development Studies, Trivandrum.
8. Ramesh, Jairam (2008), "*India's Economic Integration with Asia*", Invited lecture delivered at the Asean Development Bank (mimeo).
9. Rao MR. Palada MC. Becker BN. (2004), Medicinal and aromatic plants in agroforestry systems. Agroforestry Systems 61;07-122.
10. http://www.karvycomtrade.com/v3/Disclaimer.aspx